Pomegranate Production and Marketing

Pomegranate Production and Marketing

Editor

Andrick Cuevas

scitus
academics

Pomegranate Production and Marketing
Edited by **Andrick Cuevas**

ISBN: 978-1-68117-707-6
Library of Congress Control Number: 2016958743

SCITUS Academics LLC,
616, Corporate Way, Suite 2, 4766,
Valley Cottage, NY 10989

www.scitusacademics.com

Notice

Printed in United States of America on Acid Free Paper

Preface

Pomegranate cultivation today is a highly lucrative and remunerative agriculture business in most part of the world. The alluring monetary return per unit area from this crop has resulted in steady increase in area, production and export of pomegranate during last two decades. It is widely cultivated throughout India, Iran, China, Turkey, USA, Spain, Azerbaijan, Armenia, Afghanistan, Uzbekistan, the Middle East, Pakistan, Tunisia, Israel, dry regions of Southeast Asia, Peninsular Malaysia, the East Indies and tropical Africa. The wonderful variety of Pomegranate is known for its sweet taste, plentiful juice and health benefits that may help with Heart disease, Cancer and problem associated with Aging. Pomegranate is the fruit available throughout the year due to its increasing production. In the recently years farmers turned towards pomegranate production due to its high returns. It results in to the production of pomegranate increased greatly due to which the farmers are suffering so this text tried to analyze the problems faced by farmers during marketing phase. It is also important to provide marketing facilities to the farmers so as to improve the present conditions in the pomegranate marketing and provide better returns to the farmers.

Pomegranate Production and Marketing delivers state-of-the-art scientific and practical information about sustainable production, storage, processing, and marketing of pomegranate. It deliberates the simple and practical processing and storage techniques for the extending of shelf life of fresh squeezed 100% natural pomegranate juice while still retaining its safety and nutritional quality.

Its versatile adaptability to a wide range of climatic conditions, hardy nature, low water requirement, good response to high tech-horticultural practices high yield, returns on investment higher than many crops of dry regions, magical therapeutic values and increasing demand for table and processed products in addition to high export potential has made pomegranate a popular fruit of tropical and subtropical regions in recent periods. Though, complete know how of the crop is necessary in order to reap maximum benefits.

TABLE OF CONTENTS

1

CHAPTER

Pomegranate Processing and Value Addition: Review

Dhinesh KV* and Ramasamy D

CFDT, TANUVAS, Chennai, India

ABSTRACT

Minimally processed ready-to-eat pomegranate arils have become popular due to their convenience, high value, unique sensory characteristics, and health benefits. Pomegranate is one of the most important fruit crops in India because of its adaptable nature, high profitability and being cultivated on a commercial scale in India and the fruits are good source of nutrients and bioactive compounds, mainly anthocyanins which exhibit strong chemo-preventive activities such as anti-mutagenicity, anti-hypertension, anti-oxidative potential and reduction of liver injury. The research on the development of the new pomegranate derived products such as minimally processed pomegranate seeds, jams, marmalades, single strength juices, jellies, juice concentrates, frozen seeds, refrigerated seeds, seeds in syrup, candied arils, arils in brandy and in vinegar, carbonated beverages, pomegranate wine, pomegranate syrup etc. has been carried out. The processed products such as anardana, juice, concentrate; syrup and jelly were highly acceptable because of their nutritive and dessert qualities and palatability. The modified atmosphere packaging offered an additional innovative tool for the optimal use and value addition of lower grade pomegranate fruits.

The minimally processed pomegranate arils and frozen arils packed in punnets and pomegranate juice are the most appealing products to consumers than whole fruit. Pomegranate juice can be used in beverages, for jellies, for preparation of pomegranate juice concentrate, as flavouring and colouring agents and for dietetic and prophylactic treatment purposes. This new sector of pomegranate industrial processing will allow the use of non commercial pomegranate fruits with some physical defects and fruit disorders, having the good quality juice and seeds, to the preparation of these new products, thus improving pomegranate utilization for human health.

Keywords: Processing; Pomegranate; Utilization; MAP; Value addition; Packaging

INTRODUCTION

The pomegranate has been grown since ancient times for its delicious fruits and as an ornamental garden plant for its red, orange or occasionally, creamy yellow flowers. The pomegranate (Punica granatum L.) belongs to the punicaceae family. It is also known as the Chinese apple or Apple of Carthage or Apple with many seeds. Pomegranate is known as a super fruit of next generation and is a native of Iran to Himalayan region and is extensively grown in Iran, Spain, India and USA as well as in most Near and Far East countries [1]. In India, pomegranate is considered as a crop of the arid and semi arid regions because it withstands different soil and climate stresses. It thrives best under hot dry summer and cold winter provided irrigation facilities are available [2]. Owing to its low maintenance cost, tolerant to biotic and abiotic stresses, high yielding potential, better keeping quality and higher nutraceutical fruit value, popularity of pomegranate is increasing among the growers and consumers worldwide. It is found growing wild in the hills of Himalayas covering the entire hilly tract of Jammu and Kashmir, Himachal Pradesh and parts of Uttar Pradesh. The main pomegranate growing states in India are Maharashtra, Karnataka, Gujarat, Rajasthan, Uttar Pradesh, Andhra Pradesh and Tamil Nadu. India ranks first in the pomegranate production (8.07 lakh tones) in the world, on an area of 1.09 lakh hectares with productivity of 7.40 t/ha [3]. Maharashtra, a pomegranate basket of India, covers 0.82 lakh ha area (75%) with the production of 5.50 lakh tones (68% of the total pomegranate production in the country). More than 90 percent of the fresh produce is utilized for domestic fresh consumption and export. Spain (45%) and Iran (15%) competes the India in International market.

Excellent flavour, nutritive value and medicinal properties of pomegranate fruit indicates its good potentiality for processing into value added products having extended shelf life. In order to exploit and popularize the medicinal and nutritive values of the pomegranate to its fullest extent, it becomes essential to explore the different ways of minimal processing and post harvest technology applications. In India, in spite of the known nutraceutical benefits and great global demand for potentially pomegranate derived products; the pomegranate processing industry is not developed due to lack of technological developments for commercialization, resource personnel and scientific research database. Thus the post harvest losses are very high (20-40%). About 10- 15% fresh produce lose their market value and consumer acceptability due to improper post harvest management. Minimizing these losses can increase their supply without bringing additional land under cultivation. A number of processed products can be manufactured and preserved for future time satisfying the consumer perception of a high nutritional quality and convenience produce. These could also help to have the good returns and make the availability of the fruit throughout the year increasing the shelf life of the pomegranate fruits as such for a considerable period.

The purpose of this review paper is to present the recent advancements, current status and developments in the pomegranate processing and their utilization. It also focuses on the increasing interest in the search for new pomegranate derived food products, the application of the new post harvest technologies and identifies future prospects for the development and utilization of pomegranate processing.

Pomegranate Fruit Quality Characteristics

Pomegranate fruits are irregular round in shape with coriaceous rinds that vary from yellow, green or pink to bright deep red, depending on the variety and stage of ripening [4]. The fruit, a false berry, balausta type, is internally having multi ovule chambers separated by membranous walls (septum) and a fleshy mesocarp. The chambers are filled with shiny red seeds encased in a succulent and edible red pink pulp called arils. The arils develop from the outer epidermal cells of the seed and elongates to a very large extent in a radial direction [5]. The colour of arils varies from white to deep red depending upon the variety. Fruit quality depends largely on sugar and acid content of the juice. The edible portion of the portion of pomegranate is an excellent dietary source as it contains a significant proportion of organic acids, soluble solids, polysaccharides, vitamins, fatty acids and mineral elements of nutritional significance (Tables 1 and 2) [6,7]. The physico-chemical properties of pomegranate fruit cultivars have reported by several researchers [8-10]. They showed that the physico-chemical properties of pomegranate cultivars vary among the agro-climatic zones. Several chemical properties such as anthocyanins, vitamin C, total phenolics and total tannins have also been reported by Al-Said, et al. [11] and D'Aquino et al.[12].

Therapeutic Properties of Pomegranate

Pomegranate is a very promising and emerging crop for its refreshing arils, juice and its chemo preventive properties having medicinal value [13]. The pomegranate has been regarded as a food medicine of great importance for therapeutic purposes like colic, colitis-diarrhea, dysentery, leucorrhea, paralysis and headache [1,14]. It finds wide application in the traditional Asian medicines both in Ayurvedic and Unani systems. The therapeutic properties are reported to be due to the presence of betulic and urosolic acids and different alkaloids such as pseudo pelletierine, pelletierine and some other basic compounds [15]. There has been a remarkable increase in the commercial farming of the pomegranates globally, due to the potential health benefits of the fruit such as its high antioxidant, anti-mutagenic, anti-hypertension activities and the ability to reduce liver injury [16- 19]. Pomegranate anthocyanins have been demonstrated scavenging activities. The pomegranate polyphenolic compounds are able to elevate the antioxidant capacity of the human body. Pomegranate fruit is also known for its anti-inflammatory and anti-atherosclerotic effect activity against osteoarthritis, prostate cancer, heart disease and HIV-I [20,21]. The juice from the pomegranates is one of the nature's most powerful antioxidants. Gil, et al. [22] reported the antioxidant activity of pomegranate and compared to those of red wine and a green tea infusion showing that the commercial pomegranate juices had three times higher antioxidant activity than those of red wine and green tea. Pomegranate juice also increases the body's resistance against infections, acts as cooling beverage and tones up the function of kidney, liver and heart. All the parts of the tree, the roots, the reddish brown bark, leaves, flowers, rinds and seeds have featured in medicine for thousands of years as they are rich source of different chemical constituents (Table 3). The sweet varieties of pomegranate are considered a good laxative while those which are intermediate between sweet and sour are regarded as valuable in stomach inflammations and heart pain. The pomegranates have recently been found to boost activity of an enzyme which protects the cardiovascular risks.

Pomegranate Processing

Ongoing global drive for a healthier diet has led to a rise in demand for convenient and fresh food produce, with high nutritional value and free of additives [23]. Pomegranate indicates the great scope for the processing into value added products having extended shelf life. The fruit disorders such as sun burnt husks, splits and cracks and husk scald on whole fruit reduces marketability and consumer acceptance. The new sector of pomegranate processing allows the use of the fruits with low quality fruits that cannot be commercialized, for the preparation of the new products. Despite of great potential for pomegranate derived products, the industrial processing of pomegranate is scarce due to peeling difficulties and lack of technological development for industrial processing of pomegranate [24,25].

The activity of research and development on pomegranate has aimed at the application of new refrigerated technologies to extend the commercial shelf life of pomegranate to cash the market for export of fresh pomegranates keeping its original quality and at the search of new pomegranate derived products. The pomegranate can be processed into products like minimally processed fresh arils, juice, squash, beverage, molasses, juice concentrates, frozen seeds, jam, jelly, marmalades, grenadine, wine, seeds in syrup, pomegranate spirits, pomegranate powder, pomegranate rind powder, anardana, confectionery, pomegranate seed oil etc. [26]. Figure 1 shows the possible utilization of the pomegranate fruit in industrial processing.

New Technologies for Pomegranate Storage

The pomegranate is classified as a non-climacteric fruit. In spite of the non climacteric nature of the fruit, quantitative and qualitative loss still occur due to postharvest handling processes, resulting in chilling injuries, husk scalding, weight loss and decay of pomegranate [27]. The new physical treatment applications have been reported to prolong the shelf life of the fresh pomegranates. These treatments modifies the environmental conditions of pomegranate storage, effecting the fruit physiology and biochemistry and inhibiting the development of micro-organisms contaminating the fruit surface, keeping the original physico-chemical quality of the fruit. Kader, et al. [27] and Artes [28] recommended a fast pre-refrigeration using forced air as one of the simplest ways to extend the commercial life of pomegranate up to 2-3 months by keeping storage temperature around 5°C. Artes and Tomas- Barberan [29] reported the applications of controlled and modified atmospheres (CO2 enriched and/or reduced O2), use of the thermal treatments for fruit conditioning and curing and intermittent warming during the cold storage to avoid fungal developments and physiological disorders that develops below 5°C.

Active MAP involves a quick process of gas flushing or gas replacement or the use of gas-scavenging agents to establish a desired gas mixture within the package [30]. Studies have shown that modified atmosphere packaging (MAP) and controlled atmosphere storage (CAS) have the ability to delay quality loses and thus extends the shelf life of fresh or minimally processed or fresh-cut produce [19,31- 34]. Modified atmosphere packaging can result in reduction in the respiratory activity by decreasing O2 concentration, delay in softening and ripening and a reduced incidence of the various physiological disorders and pathogenic infestations.

MAP sensing and Monitoring: 'Smart' or 'active' or 'intelligent' packaging system is intro duced to improve the safety of MAP products and to extend the technology to a broader spectrum of products [32]. Summers [35] defined the Smart packaging as an interaction between the packing system and the product itself which confers intelligence appropriate to function and use of the product with the ability to sense or to be sensed and to communicate. Nano biosensors can serve as the best smart packaging tool for MAP sensing and monitoring. Artes, et al. [36] recommended a controlled atmosphere of 5% O2 + 0% to 5% CO2 composition for the storage of Mollar pomegranate at 5°C with RH 95% to minimize decay weight loss and chilling injuries. In contrast, Kader [37] recommended a gas composition of 3% to 5%O2 + 5% to 10% CO2 for storage of pomegranate at 5°C. Studied also reported the chilling injuries to pomegranates when stored at temperatures lower than 5°C [28,38].

Pomegranate Juice Processing

Pomegranate contains 48 to 52 per cent of edible part on the whole fruit basis, which comprises of 78 percent juice and 22 percent seed. The seeds along with arils are crushed and juice is extracted and marketed as a fresh juice due to its excellent flavour, attractive fragrance, delicious taste and high nutritive and medicinal value. Production of juice from the pomegranate arils proved to be one of the important methods of value addition. The juice can be processed possible into the squash, syrup, nectar, jelly, concentrate and such other products. Pomegranate juice can be used as an ingredient providing colour to the other products. The pomegranate juice is a rich source of polyphenols. The phenolic constituents of pomegranate such as the anthocyanins give the colour and other polyphenols such as flavonoids and some non-flavonoids are responsible of antioxidant properties, astringency and bitterness to juice [22]. The antioxidant qualities of pomegranate juice, makes it appealing for the production of health supplements and nutraceuticals [39].

Extraction of pomegranate juice

Up to date, commercial pomegranate juice production has only been mentioned by Cemeroglu [40], Vardin [41] and Saxena, et al. [2]. The foremost challenge in juice extraction is the peeling of the fruit as it is time consuming and irritating as the hands get stained due to polyphenols and oxidative enzymes. For juice extraction, the fruits are prepared by rolling it on the hard surface to weaken the seed sacks within the fruit. A prototype machine which can separate arils and skin from fruits without causing damage to arils has been developed by MPKV, Rahuri. A hand tool is designed by CIPHET, Ludhiana, India, for easy extraction of arils from fruits. The basic method for extraction of juice involves the cut opening of the fruit, seed separation and pressing in screw press or basket press. In another method, the fruits are quartered and crushed or the whole fruits may be pressed in hydraulic press and juice is strained out. Juice is extracted from the mature pomegranate seeds of sweet type cultivars like Mollar, Bhagwa etc. by pressing and liquefying. The pressing method gave the greater quantity as well as better quality. On whole fruit basis the juice yield is about 42% while on aril weight basis yield is about 70% [42]. Ozkan [43] described packaged press method for extraction of pomegranate juice from separated arils and clarified the pomegranate juice with gelatin at 4°C overnight and stored filtered juice at -30°C. While Saxena, et al. [2] recovered 36.41% the juice by cutting the pomegranate fruits into quarters and pressing them in a rack and cloth hydraulic press under moderate pressure. The hydraulic extraction of juice should be at a pressure less than 100 psi to avoid undue yield of tannins from the rind. The phenolic constituents of the pomegranate juice are responsible for the colour, astringency, bitterness as well as the formation of cloudy appearance of fruit juices during concentration and storage as reported by de Simon, et al. [44] and Spanos [45]. Adsule [46] reported the average pomegranate juice recovery by hand press method. The yield of juice was about 50 per cent on whole fruit basis, while from grains the yield was about 76 to 85 per cent. Chobe [47] separated grains and rind by pomegranate seed extractor and recovered 55 per cent pomegranate juice by crushing them in screw type juice extractor. Vardin and Fenercioglu [48] obtained the 30 to 40 percent pomegranate juice by pressing the whole pomegranate fruit in the manually operated packaged type press in batches of 10 kg for 5 minutes. They also reported that the pomegranates can be pressed as a whole or as a divided or as granulated sac, but the best results were obtained by using whole fruits considering the time and cost but resulted in the juice with excessive astringency and bitterness due to squeezing of phenolic compounds. Neifar, et al. [49] extracted pomegranate juice having pH 4.1 and TSS 150Brix by using the Philips electric juice centrifuge (centrifuge method) and stored the juice at -20°C. Alpher, et al. [50] described the process of extraction of pomegranate juice by pressing the two halved cut fruit into laboratory type press which yielded 40-50 percent juice. Marti, et al. [51] described a process for extraction of pomegranate juice having 3.8 pH, in which, aril pressurization followed by centrifugation at 6000 rpm for 10 minutes was used. Kumbhar, et al. [52] reported that the average pomegranate juice recovery on whole fruit basis by hand press method, mechanical method and screw type juice extractor method was 71.6, 66.0 and 81.0 per cent, respectively. The pomegranate juice can also be extracted by squeezing the arils gently by hand press method [53,54].

Clarification of pomegranate juice

Clarification or fining is one of the most important steps in fruit juice processing. The procedure helps to remove active haze precursors and thus decrease the potential for haze formation during storage. The rind of the pomegranate fruit contains a very large amount of excessively 'puckery' tannin which enters the juice and makes it undrinkable if the whole fruit is crushed or pressed with excessive pressure [41]. Nutritionist, however, recommend in contrast to preserve these compounds during the fruit juice processing because of their health protective effects. Pomegranate juice contains only trace amount of pectin. Therefore it can be filtered easily after pressing without clarification. However clarification is necessary to prevent the formation of cloudy appearance during storage and also to improve the taste of the product. If the clarification is not employed, the product has bitter taste due to high tannin content.

These polyphenols contribute to haze formation through the mechanisms involving prior polymerization or condensation leading to the formation of the polymeric complexes and collected at the bottom of the fruit juice bottle when stored. In pomegranate, most of the tannin presents in the outer part of the fruit. During the pressing, it passes to the juice. The main purpose of the clarification is to reduce the amount of the tannin and decrease the astringency of the product. Effective use of clarification agents requires optimization of their method of preparation as well as determination of the appropriate concentration needed to achieve clarification. For clarification, gelatin, bentonite, clays etc. may be used as flocculating agent. Sahin et al. [55] observed the positive charge of gelatine at the pH of fruit juice which removed the negatively charged phenolic compounds, by forming heavy flocculent precipitate forms. But if the gelatin is added above the level, turbidity increases [56]. Pectinase enzyme plays important role in clarification of the fruit juices by depectination [57]. The centrifugation method may also be employed for the clarification of the fruit juices. Vardin [41] reported the conventional heating treatment to raw pomegranate juice to inactivate naturally present enzymes and to destroy the vegetative microorganisms. He further opined that heating or pasteurization of the pomegranate juice can be applied after clarification and filtration as heating before clarification increases the stability of haze formation which hinders the clarification of juice permanently. The natural clarification can be suitably employed for clarification of pomegranate juice because the pomegranate juice was quite resistant to microbial spoilage at refrigeration temperature due to the presence of the polyphenolic compounds [40]. The most effective method to remove the phenolic compounds in pomegranate juice was the conventional fining with gelatin (300 mg/ l) and bentonite (300 mg/ l) along with PVPP [50] (Figure 2). Vardin and Fenercioglu [48] clarified the basket pressed juice with gelatin, poly-vinyl-poly-pyrrolidone (PVPP) and natural sedimentation and reported that the phenolic substances were controlled in each clarification method. The most effective method of clarification was the application of 1 g/l gelatin before heat treatment as it reduced the phenolic substances to an acceptable level, decrease turbidity, preserved anthocyanins and colour density. In order to reduce the amount of tannin in pomegranate juice, Bayindirli, et al. [55] found the addition of 2 g/l gelatine as the most effective method of clarification which resulted in the clear and rich coloured juice as compared to the natural clarification which gave turbid juice. Removal of phenols can be accomplished with the help of filtration methods such as ultra filtration. Neifar, et al. [49] investigated that the lactase enzyme application to pomegranate juice resulted in the 40 percent reduction of the total phenol but induced a threefold decrease of juice clarity. This drawback was overcome by the ultra-filtration of lac case treated juice giving clear and stable pomegranate juice. Pectinases play a crucial role in clarification, extraction, in reduction of viscosity, to remove off peels and to increase the yield of fruit juices. The use of pectinase enzyme for the clarification of the fruit juices by depectination is reported by many authors [57,58].

Packaging of pomegranate juice

Packaging material selection as well as processing influences the quality of foods, altering colour and nutrient composition during storage as a result of contact with oxygen and light transmission through them. Paperboard cartons with low density polyethylene (LDPE) coating or glass containers are commonly used materials of juices. Oxygen and light have destructive effect on the anthocyanin during storage. So the packaging material also plays an important role in the colour stability of stored pomegranate product. Perez-Vicente, et al. [59] assessed the influence of packaging material on colour and bioactive compounds of pasteurized pomegranate juice during storage at 24/18 °C and 40-50% RH. They opined that the organoleptic quality of juice could be altered by packaging material, even if nutritional quality is not influenced suggesting that the oxygen permeability of the packaging material (which is the more damaging factor than light for pomegranate juice) should be minimized to avoid the detrimental effects on the retention of colour and some bioactive compounds. Glass containers were found to be better as compared to high density polyethylene or polyvinylchloride containers with regards to retention of anthocyanins, vitamin C and organoleptic quality of the fruit juices [60]. Wasker and Deshmukh [61] studied the effect of light on the quality of stored pomegranate juice. The results showed that pomegranate juice packed in amber coloured glass bottle retained better anthocyanins as compared to juice packed in colourless bottles.

Storage of pomegranate juice

The numbers of factors are affecting the stability of the coloured pigments like anthocyanin in the pomegranate juice includes temperature, oxygen, light, pH, enzymatic action etc. Out of these, the storage temperature is the important one. Many researchers have reported that the pomegranate juice can be successfully stored at ambient temperature as well as in cold storage (5 ± 12°C). But the better retention of anthocyanins and reduction in the enzymatic activity of pomegranate juice was reported in the pomegranate juice stored in cold storage. Ahire [62] reported that the hand pressed or mechanical pressed pomegranate juice packed in the glass bottles could be stored satisfactorily upto 3 months under cold storage (51°C) conditions with higher acceptable organoleptic score. The pomegranate juice can be stored up to 60 days at room temperature with the minimum changes in TSS, acidity, pH, total sugar, reducing sugars and tannin when pasteurized at 70°C and added 500 ppm sodium benzoate [63]. Adam and Ongley [64] reviewed the beneficial influence of low temperature storage on various pigmented fruit products. They observed that, bottling of fruits at low pH (between 1-2) without adding sugar, led to small but significant increase in the pigment stability. Especially when stored at ambient (35°C) the pigment degradation was faster. An increasing trend in TSS, pH, reducing sugars, non-reducing sugars and total sugars during three months storage of pomegranate juice based carbonated beverage both at ambient and in cold temperature storage was reported by Shelar [65] and Sandhan [53].

Minimally Processed (Ready to Eat) Pomegranate Products

Fresh pomegranate arils

As earlier discussed, tedious, difficult and time consuming procedure in the preparation of the arils, makes the pomegranate fruit unpopular as a table fruit. It is for these reason that the development of 'ready to eat pomegranate arils has been a challenge that has been approached by several research groups in Spain Artes, et al. [66] and the USA. In recent years, minimally processed "ready-to-eat" pomegranate arils have become popular due to their convenience, high value, unique sensory characteristics and health benefits. James Caleb, et al. [23] reported that with increasing global interest in postharvest handling and nutritional value of pomegranate, MAP of minimally processed pomegranate arils offers additional tool for optimal use and value addition, including the utilization of the lower grade fruits with superficial peel defects like cracks, splits, sun burnt and bacterial blight.

Maintaining the nutritional and microbial quality of pomegranate arils is a major challenge as the minimally processed arils easily deteriorate in texture, colour, overall quality and a reduction in shelf life [19,25]. Minimal processing of pomegranate mainly consists of washing with the sanitizing agents to reduce the initial microbial load, pH modifications, use of antioxidants, modified atmosphere packaging and temperature control. (Sepulveda, et al. [34] (Figure 3) The best results with the cultivar 'Mollar de Eche' were obtained washing the arils with the chlorine solution, followed by a mixture of ascorbic and citric acids and storing the seeds at 1°C in polypropylene films that allowed the formation of a modified atmosphere appropriate for the conservation of these arils. The preparation of the arils under very clean conditions and at temperatures close to 0°C prolonged the life of this product and maintained its quality [19]. Storage at the higher temperatures (4-8°C) produced the product with lower quality and a shorter commercial life. The novel technologies such as smart packaging offers potential of increase in the shelf life of pomegranate arils by ensuring the microbial safety and monitoring the storage temperature with TTI. Also, the use of natural or non destructive products as preservatives (such as honey and UV-C radiation) should be done in combination with MAP [23]. Ayhan and Esturk [67] found that the pomegranate arils packed with air, nitrogen and enriched oxygen kept acceptable quality attributes on 18th day, however, marketability period was limited to the 15th day for the low oxygen atmosphere. Lopez-Rubira, et al. [24] studied the effect of harvest time, use of different UV-C radiation and passive MAP with polypropylene basket sealed with BOPP film storage on sensory, chemical and microbial quality as well as on the shelf life of minimally fresh processed pomegranate var. Mollar de Elche arils. They observed that the rate of respiration of fresh processed arils was higher in the late harvest than in the earlier harvested fruits. No significant differences were observed between the control and UV-C treated arils and there was no observable interaction between passive MAP and UV-C treatment, except that the CO2 accumulation within aril packages was higher in December harvest than those of October, due to their respiration rate. They reported that unclear results were obtained on the effect of UV-C radiation on the microbial growth of pomegranate arils and the shelf-life of commercially produced pomegranate arils can be increased up to 10 days with use 100% nitrogen in pet packages. No significant change in the total anthocyanin content of arils was observed. They also reported that the microbial counts of minimally fresh processed arilsincreased throughout the shelf life, with mesophilic counts of control arils processed in October slightly higher than those from December. A low count of micro-aerophilic lactic acid bacteria after 10 days of aril storage without any trace of fermentative metabolism was also recorded. Nunes, et al. [68] investigated the metabolic response of the minimally processed pomegranate arils cv. Hicanar to UV-C treatments and stored at 2°C and 6°C. They observed that the UV-C illumination had an effect on the increase of phenol content in pomegranate arils but TSS and citric acid percentage remained unaffected.

Gil, et al. [25] reported that the pomegranate arils of Mollar variety washed with chlorinated water (100 mg/kg) and antioxidants solution (5 g/l ascorbic acid and 5 g/l citric acid), packed in OPP film, using an initial atmosphere actively modified to 0 ml/l CO2 and 20 ml/l O2 and stored at 1°C, can be stored up to 7 days maintaining a good quality and appearance, without visible attack of moulds or off flavour developments. Sepulveda, et al. [34] investigated that pomegranate arils cv. Espanola washed and immersed in antioxidant solution (sodium hypochlorite, 200 ppm, 5% ascorbic acid and 5% citric acid) for 1 minute and stored at 5 ± 0.5°C in three semi-permeable packaging can be stored for 7 days with commercial marketable acceptance. BB4 bags showed the best ability to maintain the physical, chemical and microbiological characteristics of the minimally processed pomegranate arils, but the sensory characteristics decreased from 7 to 14 days storage. Palma, et al. [69] during his studies on the chemical and organoleptic characteristics of minimally processed seeds of pomegranate cv. Primosole, packed in a 40 μm thick polypropylene film, stored at 5°C for 10 days observed that a passive modified atmosphere was established within the package, with a progressive increase in CO2 and decrease in O2 level, with increased ethylene concentration at the end of the storage. No significant changes in chemical properties were recorded, but increase in titratable acidity was observed in packaged seeds. Hess-Pierce and Kader [70] concluded that arils of 'Wonderful' pomegranate can be stored upto 16 days at 5°C with 20 per cent gas composition of CO2 without changes in the physical and chemical characteristics of the fruit. Further they also observed the susceptibility of the mechanically damaged arils to moulds after 12 days. Ersan, et al. [71] studied the effect of CO2 and O2 on the respiration rate of pomegranate arils cv. Hicaz and evaluated the suitability of the common packaging material for designing the MAP. They reported that the respiration rate of the arils was significantly affected by O2 and CO2 concentrations in the surrounding atmosphere. Lower respiration rate was achieved in 2 per cent O2 with 10 or 20 per cent CO2 but MAP design analysis indicated that these target levels of O2 and CO2 cannot be achieved by LDPE and PP materials for commercial packages. The storage atmosphere enriched in CO2 (10, 15 and 20%) helped to prolong the commercial life of pomegranate seeds upto 8 days at 100°C and 12 days at 50°C [29]. The use of honey treatments has also been explored in preserving the fresh-like quality of arils, other cut fruits and to extend their shelf life. Honey has been used since ancient times as a sweetening agent in food and is only the concentrated form of sugar available worldwide. Ergun and Ergun [72] evaluated the efficacy of varying concentration of 10 and 20 per cent honey dip treatment on the quality and shelf life of minimally processed pomegranate arils cv. Hicaznar stored at 40°C in loosely closed containers. They reported that honey solution (10 and 20 per cent) dipping for five minutes can help in storage of arils at 40°C for 10 days with brilliant aroma than those treated with water. It also helped in the lowering the softening of the arils. The total aerobic count was lower in honey treated arils compared with the control but the counts increased across all the treatments compared with the count immediately after treatment. It extended the fresh-like quality of arils by delaying quality loss, microbial development and pigment changes, thus providing the safe organic method. A summary of MAP of pomegranate arils, different cultivars, highlighting the types of the packaging adopted and the modified atmosphere condition attained in the packages is reviewed in Table4.

Frozen arils

After preparing the arils same as that of minimally processed arils, the arils were packed in the polyethylene bags with syrup of 15°Brix. This concentration is similar to that of the arils. The arils were then frozen in a chest freezer. The juice contain of aril enters the syrup during freezing. Maestre, et al. [73] also reported the freezing of the arils coated with the sugar. The coated sugar turned red during freezing and storage. The arils should be eaten frozen to avoid an excessive loss of turgence.

Appertised arils

Appertised arils are the product prepared by putting the arils in syrup of 15°Brix and packed into metal tin. The tins were heated, sealed and sterilized for 10 minutes. After stabilization, most of the tins had arils that were too soft and that tasted cooked. The tins which were prone to less severe heat treatment had adequate textured arils with good taste.

Arils in vinegar

The arils were preserved in vinegar with an acidity of 5% and packed in jar. The resultant product was the brown coloured pomegranate arils.

Jams and preserves (Anar Rub)

A product known as anar rub with fairly good keeping quality can be made be concentrating pomegranate juice and heating the mixture on a slow fire for long period. The finished product has a thick consistency and contains 70-75% TSS [74]. It can be stored for one year and utilized as a jam. Maestre, et al. [73] reported that the jams and preserves made from the frozen Mollar pomegranate juice by adding pectins, saccharose and citric acid. It was observed that during the processing treatment 25% of the pigments are destroyed. They continue to degrade during preservation, depending on the temperature than on light. The best preservation temperature reported was 5°C.

Jellies

An attractive jelly can be prepared from pomegranate juice [42]. Adsule, et al. [75] prepared good pomegranate jelly on a small scale from Ganesh cultivar of pomegranate. While making jellies, approximately 50% of the anthocyanins present in pomegranate juice were lost. Maestre, et al. [73] investigated that the acidification of juice produced a noteworthy improvement in the colour of jelly, both initially and during storage. During storage, certain colour differences were observed, which indicates that the pH was not only the parameter responsible for this characteristic.

Anardana

Anardana is dried arils from sour wild types. It is mainly used as acidulent in place of tamarind of dried green mango (Amchur) in North India in Indian styled curries, chutney and other culinary preparations. It is also used in the preparation of digestive candies and by traditional system of Ayurvedic and Unnani medicine. The improved processing technique consists of pre cleaning, mechanized extraction of arils, solar/ sun drying and packaging. After trating with sodium benzoate (600 ppm) for 10 minutes, arils are dehydrated in a drier at 45°C for 48 hrs to 10-12% moisture content. It has attractive brown colour and can be stored for a long time in glass jars [76].

Pomegranate molasses

It is traditional Middle Eastern ingredient made from cooked down pomegranate juice. Thick and syrupy in texture, pomegranate molasses provide tangy flavor and is dark in colour. Its sweetness comes from the concentration of the fruits natural sugars [26]. To make ½ cup of molasses, 4 cups of juice is heated in a pan for 45 minutes, allowing it to thicken but not overcooked. The product can be stored in air tight container under refrigerated conditions for 3 months. It is typically used to flavor chutneys, curries and salad dressings to glaze or tenderize meat products.

Pomegranate concentrate

The commercially available pomegranate juice concentrate generally contains 65-70% TSS and pH 2.7 to 3.1. It is made from natural pomegranate juice and it is generally free from any added sugar or preservative. Most of the companies in international market are promoting it as health drink giving stress on the medicinal and health benefits of pomegranate. Maskan [77] described the different processes for the preparation of pomegranate concentrate of 60.5°Brix by using microwave, rotary vacuum and atmospheric heating processes method in 23, 108 and 190 min respectively. Hulya Orak [78] recorded increase in the reducing sugars, glucose and fructose level to 46.46%, 23.89% and 22.53%, respectively during the conventional method of pomegranate juice concentration process. Also the potassium and magnesium mineral contents of concentrate increased during concentration.

Pomegranate syrup/grenadine

Pomegranate syrup is sold commercially as grenadine and is used as flavoring in alcoholic drinks, soft drinks and confections. Grenadine is light pomegranate syrup prepared by mixing juice and sugar. A syrup of about 60°Brix with an added acidity of 1.5% as citric acid has a bright purplish red colour and a delightful taste and flavor. It can be preserved by pasteurization or by addition of sodium benzoate.

Pomegranate beverages

Pomegranate can be processed into delicately flavoured RTS beverage, blend and squash. Squash is prepared with 40% juice and RTS beverage and beverage blends with 15% juice content. Squash is ready to serve in the ratio of 1:4 (one part of squash with 4 parts of water).

Confectionery

Phadnis [42] reported preparation of jelly and hard candy from pomegranate. Researchers have found that approx. 50% of total anthocyanin present in the juice is lost during jelly preparation and the acidification of juice improves the colour in the final product. Anar Rub (Pomegranate jam) is prepared from juice by adding sugar and heating to a thick consistency.

Wine

Pomegranate juice can be utilized to make good quality wine. Sugar is added to adjust Brix to 22-23°. The pasteurized juice is fermented with starter wine yeast. The fermentation is allowed to continue until desired level of alcohol is obtained. The wine is clarified by bentonite treatment or by centrifugation. The wine aged in the same manner as red grape wine.

Pomegranate powder

Low moisture pomegranate powder is prepared from pomegranate arils that have been dried and milled. The active ingredient considered in pomegranate powder is ellagic acid. The freeze dried pomegranate powder rich in ellagic acid have many applications and may be encapsulated for regular doses [79]. Husain, et al. [80] have proposed grade standards to bring the commodity under the purview of AGMARK certification.

Pomegranate Waste Utilization

All parts of the pomegranate tree i.e. roots; bark, leaves, flowers, rind and seeds can be processed for value added products having medicinal, industrial and cosmetic value. Technology for preparation of rind powder has been developed, having potential uses in medicine, leather and dye industry and in tooth powder preparation. The recovery of rind powder has been found to be 15.5% on whole fruit basis and 34% on rind weight basis.

As a part of pomegranate industrialization, the field of pomegranate by-products is an interesting field to be exploited as it is a rich source of alkaloids, aromatic compounds and enzymes.

CONCLUSION

The pomegranate fruit is considered as the suitable fruit for the processing and utilization due to its excellent flavour, colour, physicochemical constitution and therapeutic properties. It is referred as the 'Super fruit' due its high nutritional value, high antioxidant capacity and consumer appeal. But in spite of high nutraceutical value, the consumption of pomegranate is not wide spread due to tedious and time consuming peeling and irritations and staining of hands during handling of fruits. The pomegranate processing and product diversification has played important role in the increased consumption and utilization of pomegranate. Last few years, the research and development activities on pomegranate fruit have aimed to develop technologies for new pomegranate derived food products. The pomegranate can be processed into products like minimally processed fresh arils, juice, squash, beverage, molasses, juice concentrates, frozen seeds, jam, jelly, marmalades, grenadine, wine, seeds in syrup, pomegranate spirits, pomegranate powder, pomegranate rind powder, anardana, confectionery, pomegranate seed oil etc. These products are not yet popularized in large scale due to lack of the commercially viable processing technologies. The keen and immediate attention is required in meeting the research and developmental gaps for the commercialization and popularization of pomegranate processing technology and pomegranate products. The major aspects which requires immediate attention in pomegranate processing includes industrial method of peeling, standardization of extraction and proper clarification methods of pomegranate juice, development of standards for packaging and storage of pomegranate derived products, application of the new inline technologies such as MAP, ultra filtration for pomegranate utilization and popularization of pomegranate based products. Hence, for the commercialization and utilization of pomegranate processing technology and pomegranate products, experimental studies should be carried out with more informative output on the metabolic properties of pomegranate and derived products under various conditions in order to develop the scientific database and to enable the successful application of the available technology.

REFERENCES

1. Schubert SY, Lansky EP,Neeman I (1999) Antioxidant and eicosanoidenzymeinhibitionproperties of pomegranateseedoil and fermentedjuiceflavonoids. Journal of Ethnopharmacology 66: 11-17.

2. Saxena AK, Manan JK,Berry SK (1987) Pomegranate postharvesttechnologychemistry and processing. IndianFoodPacker 41: 43-60.

3. NHBMA (2010) IndianHorticulture Database-2009. NationalHorticultureBoardMinistry of AgricultureGovt of India.

4. Holland D, Hatib K,Bar-Ya'akov I (2009) Pomegranate: botanyhorticulturebreeding. In: Journal of Janick (Ed.) Horticultural Reveiws 35: 127-191.

5. Fahan A (1976) Theseed.PlantanatomyHakkibutzhameuhadPublicationJerusaleum: 419-430.

6. Ewaida EH (1987) Nutrientcomposition ofTaifipomegranate (Punicagranatum L.) fragments and theirsuitabilityfortheproduction of jam. PersianGulfScienceResearchAgricultural and BiologicalScience 3: 367-378.

7. Fadavi A, Barzegar M, Aziz H (2006) Determination of fattyacids and total lipidcontent in oilseed of 25 pomegranatevarietiesgrown in Iran. Journal of FoodComposition and Analysis 19: 676-680.

8. Artes F, Tomas-Barberan FA (2000) Postharvesttechnologicaltreatments of pomegranate and preparation of derivedproductsCIHEAM-IAMZ: 199-204.

9. Opara LU, Al-Ani MR, Al-Shuaibi YS (2009) Physico-chemicalproperties, vitamin C content and antimicrobialproperties of pomegranatefruit (Punicagranatum L.).Food and BioprocessTechnology 2: 315-321.

10. Zarei M, Azizi M, Bashiri-Sadr Z (2010) Studiesonphysic-chemicalproperties and bioactivecompounds of sixpomegranatecultivarsgrown in Iran. Journal of FoodTechnology 8: 112-117.

11. Al-Said FA, Opara LU, Al-Yahyai RA (2009) Physico-chemical and textural qualityattributes of pomegranatecultivars (Punicagranatum L.) grown in theSultanate of Oman. Journal of FoodEngineering 90: 129-134.

12. D' Aquino S, Palma A, Schirra M, Continella A, Tribulato E, et al. (2010) Influence of film wrapping and fludioxonilapplicationonquality of pomegranatefruit. PostharvestBiology and Technology 55: 121-128.

13. Hertog MGL, Van Popel G, Verhoeven D (1997) Potentiallyanticarcinogenicsecondarymetabolitesfromfruits

and vegetables. Phytochemistry of fruits and vegetables. ClaredonPress, Oxford, UK: 313-329.

14. Sadeghi N, Jannat B, Oveisi MR, Hajimahmoodi M,Photovat M, et al. (2009) Antioxidantactivity of Iranianpomegranate (Punicagranatum L.) seedextracts. Journal of AgricultureScience and Technology 11: 633-638.

15. Singh RP, Gupta AK, Bhatia AK (1990) Utilization of wild pomegranate in northwestHimalayas-Status and Problems. Proceedings of NationalSeminaronProduction and Marketing of IndigenousFruits New Delhi: 100-107.

16. Du CT, Wang PL, Francis FJ (1975) Anthocyanins of Pomegranate(PunicagranatumL.).Journal of FoodScience 40: 417-418.

17. Tsuda T, Watanabe M, Ohshima K, Norinobu S, Choi S, et al. (1994) Antioxidantactivity of theanthocuaninpigmentscyaniding 3-O-ß-D glucoside and cyaniding. Journal of Agriculture and FoodChemistry 42: 2407-2410.

18. Lansky E, Hubert S, Neeman I (1998) Pharmacological and therapeuticalproperties of pomegranates. Proceedings of thefirst International SymposiumonPomegranateSpain: 231-235.

19. Gil MI, Artes F, Tomas-Barberan FA (1996a) Minimalprocessing and modifiedatmospherepackagingeffectsonpigmentation of pomegranateseeds. Journal of FoodScience 61: 161-164.

20. Malik A, Afaq F, Arfaraz S, Adhami VM, Syed DN (2005) Pomegranatejuiceforchemoprevention and chemotherapy of prostatecancer. Proceeding of theNationalAcademy of Sciences 102: 14813-14818.

21. Sumner MD, Elliott-Eller M, Weidner G, Daubenmier JJ, Chew MH, Marlin R (2005) Effects of pomegranatejuiceconsumptiononmyocardialperfusion in patientswithcoronaryheartdisease. The American Journal of Cardiology 96: 810-814.

22. Gil MI, Tomas-Barberan FA, Hess-Pierce B, Holcroft DM,Kader AA (2000) Antioxidantactivity of pomegranatejuice and itsrelationshipwithphenoliccomposition and processing. Journal of Agricultural and FoodChemistry 48: 4581-4589.

23. James Caleb O, Opara UL, Witthuhn CR (2011) Modifiedatmospherepackaging of pomegranatefruit and arils: A review. FoodBioprocessTechnologyPublished Online bySpringerScience Business Media LLC.

24. Lopez-Rubira V, Conesa A, Allende A,Artes F (2005) Shelf-life and overallquality of minimallyprocessedpomegranatearilsmodifiedatmospherepackaged and treatedwith UV-C. PostharvestBiology and Technology 37: 174-185.

25. Artes F, Gil MI,Martinez JA (1995) Procedimiento para la conservation de semillas de granada en fresco.

26. Yadav K, Sarkar BC, Kumar P (2006) PomegranateRecentdevelopments in HarvestingProcessing and Utilization. IndianFoodIndustry 25: 55-62.

27. Kader AA, Chordas A,Elyatem S (1984) Response of pomegranates to ethylene treatment and storage temperature. California Agriculture July-August: 14-15.

28. Artes F (1992) Factores de calidad y conservacion frigorifica de la Granada. II Jornadas Nacionales del Granado. Universities Politecnica de Valencia Valencia.

29. Adam JB,Ongley MH (1972)Changes in polyphenols of red fruitsduringheatprocessing. Thedegradation of anthocyanins in cannedfruitsTech Bull 23,theCampdenFoodPreservationResearchAssociationChippingCampden Gloucestershire.

30. KaderAA,Watkins CB (2000) Modifiedatmospherepackagingtoward 2000 and beyond. HortTechnology 10: 483-486.

31. Lee L, Arul J, Lencki R, Castaigne F (1995) A reviewonmodifiedatmospherepackaging and preservation of fruits and vegetables: physiologicalbasis and practicalaspects-part 1. PackagingTechnologyScience 8: 315-331.

32. Church N (1994) Developments in modifiedatmospherepackaging and relatedtechnologies. Trends in FoodScience and Technology 5: 345-352.

33. Holcroft DM, Gil MI, Kader AA (1998) Effect of carbondioxideonanthocyaninsphenylalanineammonialyase and glucosyltransferase in thearils of storedpomegranates. Journal of American Society of Horticultural Science 123:136-140.

34. Sepulveda E, Galletti L, Saenz C, Tapia M (2000) Minimalprocessing of pomegranatevar. Wonderful. Symposiumonproductionprocessing and marketing of pomegranate in theMediterraneanregion: advances in research and technology, ZaragosaSpain: 237-242.

35. Summers L (1992)Intelligentpackaging. Centre forExploitation of Science and Technology, London.

36. Artes F, Marin JG, Martinez JA (1998) Permeabilityrates of films formodifiedatmospherepackaging of respiringfoods. FoodqualityLeuven: EuropeanCommision: 153-157.

37. Kader AA (1995) Regulation of fruitsphysiologybycontrolled and modifiedatmosphere. Acta Horticultural 398: 59-70.

38. Kader AA (1986) Biochemical and physiologicalbasisforeffects of controlled and modifiedatmospheresonfruits and vegetables. FoodTechnology 40: 99-104.

39. Singh RP, Murthy KNC,Jayaprakasha GK (2002) Studiesontheantioxidantactivity of pomegranate (Punicagranatum L.) peel and seedextractsusing in vitro models. Journal of Agriculture and FoodChemistry 50: 81-86.

40. Cemeroglu B (1977) NarSuyuOretimTeknolojisiUzerineBir Ara-stirma. Ankara UniversityPress, ZiraatFak, Yayinlari Ankara Publ 664: 8-28.

41. Vardin (2000) Theusingpossibilities of differentpomegranatecultivarsgrown in Harranplainforusing in foodindustryTurkey.

42. Phadnis NA (1974) Pomegranatefordessert and juice. IndianHorticulture 19: 9.

43. Ozkan M (2002) Degradation of anthocyanins in sour cherry and pomegranatejuicesbyhydrogenperoxide in thepresence of addedascorbicacid. FoodChemistry 78: 499-504.

44. de Simon BF, Perez-Ilzarbe J, Hernandez T, Gomez-Cordoves C, Estrella I (1992) Importance of phenoliccompoundsforthecharacterization of fruitjuices. Journal and Agricultural and FoodChemistry 38: 1565-1571.

45. Spanos GA,Wrolstad RE (1992) Phenolics of applepear and white grape juices and theirchangeswithprocessing and storage-A review. Journal of Agriculture and FoodChemistry 40: 1478-1487.

46. Adsule RN, PatilNB(1995) Pomegranate Hand bookof fruitScience and Technology Marcel Dekker. New York: 455-464.

47. Chobe RS (1999) Studiesonextractionclarificationpreservation and storage of pomegranate (Punicagranatum L.) juice.

48. Vardin H,Fenercioglu H (2009) Studyonthedevelopment of pomegranatejuiceprocessingtechnology: The pressing of pomegranatefruit . Acta Horticulture 818: 373-381.

49. Neifar M, Ellouze-Ghorbel, Kamoun A, Baklouti S, Mokni A, Jaouani A, Ellouze-Chaabouni S, et al. (2009) Effectiveclarification of pomegranatejuiceusinglaccasetreatmentoptimizedby response surfacemethodologyfollowedbyultrafiltration. Journal of FoodProcessEngineering34: 1199-1219.

50. Alper N, SavasBahceci K, Acar J (2005) Influence of processing and pasteurizationoncolourvalues and total phenoliccompounds of pomegranatejuice. Journal of FoodProcessing and Preservation29: 357-368.

51. Marti N, Perez-Vicente A, Garcia-Viguera C (2001) Influence of storagetemperature and ascorbicacidadditiononpomegranatejuice. Journal of ScienceFood and Agriculture 82: 217-221.

52. Kumbhar SC, Kotecha PM, Kadam SS (2002) Effect of method of juiceextractiononquality of pomegranatewine. Beverage and FoodWorld 29: 35-36.

53. Sandhan VS (2008) Preparation of carbonatedbeveragefrompomegranate(PunicagranatumL.) 3: 66-69.

54. Singh J, Singh AK, Singh HK (2005) Preparation and preservation of pomegranate (Punicagranatum L.) RTS. Beverage and FoodWorld 32: 45-46.

55. Sahin S, Bayindirli L,Artik N (1992) Theeffect of depectinization and clarificationon sour cherryjuicequality. Gida 17: 35-42.

56. Cemeroglu B (1982) MeyveSuyuUretimTeknolojisiTeknikBasimSanayii Ankara.

57. Vilquez F, Laetreto C, Cooke RD (1981) A study of theproduction of clarified banana juiceusingpectinolyticenzymes. Journal of FoodTechnology 16: 115-125.

58. Song-nian G, Chil X, Quing D, Ying Z (2011) Optimization of enzymaticclarification of pomegranatejuiceusing response surfacemethodology. Storage and Process 2: 1-3.

59. Perez-Vicente A, Serrano P, Abellan P, Garcia-Viguera C (2004) Influence of packaging material onpomegranatejuicecolour and bioactivecompoundsduringstorage. Journal of Science and FoodAgriculture 84: 639-644.

60. Sethi V (1985) Suitability of differentpackaging material forstoringfruitjuices and intermediate preserves. Souvenir of SymposiumonRecentDevelopment in FoodPackagingMysore India.

61. Waskar DP, Deshmukh AN (1995) Effect of packagingcontainersontheretention of anthocyanins of pomegranatejuice. IndianFoodPacker 49: 5-8.

62. Ahire DB (2007) Studiesonextractionpackaging and storage of pomegranate (Punicagranatum L.) juice.

63. Suryawanshi AB, Kirad KS, Phad GN,Patil SB (2008) Standardization of preservationmethod and theircombinationforsafestorage of pomegranatejuice at roomtemperature. TheAsianJournal of Horticulture 3: 395-399.

64. Adsule RN, Kotecha PM,Kadam SS (1992)Preparation of winefrompomegranate. Beverage and FoodWorld 19: 13-14.

65. Shelar YV (2001) Preparation of carbonatedready-to-serve (RTS) beveragefrompomegranate (Punicagramaum L.) juice.

66. FICCI (2005) Anardana In TechnologyOptions. FICCI AgribuisinessInformation Centre.

67. Ayhan Z,Esturk O (2009) Overallquality and shelflife of minimallyprocessed and modifiedatmospherepackagedReady-to-Eatpomegranatearils. Journal of FoodScience 74: 399 - 405.

68. Nunes C, Graca A, Yildirim I, Sahin G, Erkan M, et al. (2010) Metabolic response to UV-C treatment on minimally processed pomegranate arils. 6th International Postharvest Symposium. Acta Hort 877: 657-662.

69. Palma A, Schirra D'Aquino S, La Malfa S, Continella G (2009) Chemical properties changes in pomegranate seeds packaged in polypropylene trays. Acta Horticulture 818: 323-329.

70. Hess-Pierce B, Kader A (1997) Carbondioxideenrichedatmospheresextendpostharvestlife of pomegranatearils. Seventhinternationalcontrolledatmosphereresearchconference. Proceedingsfresh-cutfruits and vegetables and MAP, Dept of Pomology, University of California 5: 122.

71. Ersan S, Gunes G,Zor AO (2010) Respiration rate of pomegranate arils as affected by O2 and CO2 and design of modified atmosphere packaging. ActaHort876:189-196.

72. Ergun M, Ergun N (2009) Maintaining quality of minimally processed pomegranate arils by honey treatments. British Food Journal 111: 396-406.

73. Maestre J, Melgarejo P, Tomas-Barberan FA,Garcia-Viguera C (2000) New foodproductsderivedfrompomegranate. Symposiumonproductionprocessing and marketing of pomegranate in theMediterraneanregion: advances in research and technology, Spain: 243-245.

74. Siddappa GS, Bhatia BS (1954) The identification of sugars in fruits by chromatography. Indian Journal of Horticulture 11:19.

75. Chavan UD, Adsule RN, Kadam SS (1995) Physico-chemicalproperties of pomegranaterindpowder. Beverage and FoodWorld 22: 36.

76. Artes F, Tudela JA,Villaescua R (2000)Thermalpostharvesttreatmentforimprovingpomegranatequality and shelflife. PostharvestBiology and Technology 18: 245-251.

77. Maskan M (2006) Production of pomegranate (Punicagranatum L.) juiceconcentratebyvariousheatingmethods: Colourdegradation and kinetics. Journal of FoodEngineering 72: 218-224.

78. Bayinderli L, Sahin S,Artik N (1994) Theeffects of clarificationmethodsonpomegranatejuicequality. FruitProcessing 9: 267-270.

79. Donald EP (2005) Ingredients. FoodTechnology 59: 46-55.

80. Husain MI, Devikar GY, Singh G,Jaiswal PK (2004) Qualityevaluation of pomegranateseed and powder. Spice India 17:28-32.

2

CHAPTER

Sustainable Climate Change Adaptations in Developing Countries: A Case Study of the Turkish Cypriot Community's Adoption of Pomegranate Farming

Clare M. Finnegan, Omer Gokcekus

School of Diplomacy and International Relations, Seton Hall University, South Orange, NJ 07039, USA

ABSTRACT

Climate change is a growing issue for developing countries, as they typically lack the technical and financial inputs to implement the necessary agricultural adaptations. These countries also suffer from the classic collective action problem; although they are able to identify the issue and a potential solution, their individual resources are not substantial enough to enact change. This article discusses north Cyprus' 2006 adoption of pomegranate production and its relationship to climate-related agricultural concerns. We argue that the Turkish Cypriot community would not have been able to start an effective pomegranate agribusiness without third-party financial and technical assistance. As a post-conflict developing community, they lacked the resources necessary to collectivize on their own and initiate crop switching. Thus, Turkish Cypriot farmers needed external resources in order to launch a sustainable development project. The programme was a successful example of sustainable peacebuilding as it required local ownership.

Keywords: Cyprus Pomegranate Agribusiness Climate change Crop switching Post-conflict Sustainable peacebuilding

1 INTRODUCTION

Climate change used to be an abstract, futuristic problem. Although there was some awareness of the need to adapt one's lifestyle to be more "environmentally friendly", the problem lacked urgency and was perceived as an issue that the coming generation would tackle. Now it seems to be an inescapable reality. Developed countries with abundant resources, like the USA, have been able to implement certain adaptations to mitigate some of the adverse effects of climate change. How have other less affluent countries adapted?

As what Gokcekus and Bengyak (2014) discuss in regard to corruption, climate change has the most severe ill effects on those members of society (or the world) that are least capable of preparing themselves for its impact. Often still largely agriculturally based, the economies of many developing countries are especially vulnerable to environmental changes. Limited financial resources create a collective action problem; while the individual members of a developing society may know that the adaptation or modification of existing processes can help combat climate-related issues, they lack the resources—technical and financial—to adopt these changes. The problem of adaptation is particularly acute in post-conflict regions that have been isolated from beneficial economic relationships, such as north Cyprus.

North Cyprus has been a de facto separate country since Cyprus was partitioned by the "Green Line"—a UN buffer zone—in 1974 due to conflict between Cypriots of Greek ancestry and those of Turkish

(BBC 2011). Under the internationally recognized Republic of Cyprus, Greek Cypriots were able to interface with the EU and the greater international community. North Cyprus was isolated, and its economic development has lagged behind that of the Republic of Cyprus. On average, Greek Cypriots have a 61 % higher per capita income than Turkish Cypriots (Gokcekus 2008, p. 15). This paper will explore how intervention by a third party helped the Turkish Cypriot community (TCC) adapt its agricultural industry to the constraints of climate change and simultaneously overcome its collective action problem. Specifically, this paper explores USAID's 2006 agribusiness programme that assisted the TCC with starting commercial production of the nutritional superfruit pomegranate.

2 CROP SWITCHING AND SUSTAINABLE PEACEBUILDING

Why does climate change matter? Kurukulasuriya and Mendelsohn (2008) try to predict the impact of climate change on farms' output and revenue in 11 African countries. Under one climate change scenario—the Canadian Climate Centre (CCC) model in which the studied regions become hotter and drier—predicted revenues for the year 2100 would decline by nearly 70 % if farmers did not participate in crop switching (p. 122). In an alternative climate change scenario—the Parallel Climate Model (PCM) in which the climate of the studied regions becomes mild and wet—revenues were actually expected to significantly (82 %) increase. Which of these two models most closely predicts future conditions is currently unknown. In order to avoid a potential 70 % loss of revenue, the authors suggest crop switching; under the CCC's conditions, crop switching could reduce losses to only 6 %.

Seo and Mendelsohn (2008) also explored crop choice and its relationship to climate change, focusing on South American farms. As in Africa, crop choice in South America generally had an identifiable, significant relationship with temperature and precipitation. The one exception to this was maize, a crop with many varieties that are adaptable to most South American climates. Seo and Mendelsohn (2008) did not predict the economic impact of not crop switching, but instead assumed that farmers—making a rational economic decision—would switch.

Huda et al. (2005) suggest measures that farmers, community workers, and policy agencies can take in response to changing environmental conditions. Huda et al. were primarily concerned with the possible mismatch between crop phenology and water availability caused by climate change. Without an understanding of phenology, crop adaptation proves difficult as crops with certain phenologies only thrive in specific ecosystems. Huda et al. (2005) therefore recommend several strategies, such as changing the time of planting and adopting alternative crops, that could help mitigate mismatch between crop phenology and water resources (p. 142).

Farmers can only combat climate change by employing crop switching if "there are no barriers to the adoption of appropriate crops" (Kurukulasuriya and Mendelsohn 2008, p. 123). Seo and Mendelsohn (2008) echoed this warning, specifically mentioning that farmers may not be able to crop switch "if the adjustment requires a heavy capital investment" (p. 115). Lobell et al. (2008) argue that while "switching to an existing crop variety may moderate negative impacts, the biggest benefits will likely result from more costly measures…" (p. 607). In addition to potential financial constraints, the measures advocated by Huda et al. (2005) generally require significant technical or scientific acumen, knowledge that is often in short supply in developing nations.

These financial and technical constraints only exacerbate the collective action problem, which is particularly acute when resources are scarce. In a society with limited inputs, the participation of many actors (a group effort) is often necessary in order to secure opportunities or goods that enhance the collective benefit. As Olson (1965) details, large groups are particularly prone towards free-ridership, since members of the group believe their non-participation will be overlooked. This tendency is especially detrimental to projects in developing countries; the constrained resources of individual members in these countries often require that collective goods be provided through larger-sized groups with high levels of participation.

How can developing countries, particularly post-conflict societies like Cyprus, solve these challenges? Gokcekus et al. (2014) suggest that the solution lies with intervention by an outside actor. In their study of Turkish Cypriot beekeepers, Gokcekus et al. found that the beekeepers needed EU financial and technical assistance to overcome their collective action problem, meet the EU's health standards for commercial honey, and initiate trade across the Green Line. The EU provided the initial impetus for change; once the Turkish Cypriots beekeepers understood the standard requirements for trade and had the financial resources to meet these standards, they were able to collectivize and increase their level of economic development. In other words, although the process of updating their beekeeping practices to meet EU standards was instigated by the EU, Turkish Cypriot beekeepers accepted local ownership of the remainder of the development process.

As Donais (2009) notes, local ownership is relatively rare in post-conflict development initiatives, since the external actors tend to reserve "most key decision-making authority" (p. 4). Perhaps the interaction between the beekeepers and the EU can best be described as an example of what Donais terms "sustainable peacebuilding" (p. 14). Sustainable peacebuilding requires "negotiated hybridity" between both insiders and outsiders "in which the division of responsibilities between outsider and insider is constantly calibrated and adjusted as a means to advancing the peace process" (p. 14, 21). Like the case of the beekeepers, the decision of Turkish Cypriot farmers to commence pomegranate production is likely another example of sustainable peacebuilding.

The case of pomegranate production in north Cyprus can be approached from several perspectives. This paper concentrates on intersection of two of those: (1) the adoption of pomegranates as a form of crop switching and (2) the adoption of pomegranates as a result of overcoming a collective action problem through sustainable peacebuilding. By promoting local ownership and sharing responsibility, an external actor helped the TCC surpass the capital and technical barriers that had previously prevented the implementation of a climate change solution.

3 CLIMATE CONSIDERATIONS IN CYPRUS

Climate change has exacerbated Cyprus' water shortages (Shoukri and Zachariadis 2012; Zachariadis 2010; EEA 2009) Cyprus has experienced extended periods of drought since the 1970s; increased demand has made the issue of water availability particularly acute (BBC 2010). In an attempt to rectify this issue, north Cyprus agreed to the construction of an underwater freshwater pipeline connected to Turkey. Referred to as Barış Su ("Peace Water"), this pipeline was expected to be completed sometime in 2014; however, it was slightly behind schedule and just over halfway completed in October of that year (Gies 2013; Sands 2014). In addition to political concerns over heightening north Cyprus' reliance on Turkey, there are also worries that the pipeline will only increase demand (Gies 2013). Water shortages have been particularly challenging for the agricultural sector. Citrus fruits, Cyprus' third most lucrative agricultural export, place considerable stress upon Cyprus' limited water resources; they originated in the tropics and have the greatest water consumption of Cyprus' irrigated crops (Sofroniou and Bishop 2014; Dworak and Berglund 2012). With Cyprus' continuing water limitations, relying on water-intensive crops like citrus fruits to drive the economy is not a sustainable development strategy.

The TCC's adoption of pomegranate farming was a response to climate change. Cyprus' competitive edge in the citrus market had been undermined by countries that do not share its climate constraints; farmers needed to implement crop switching to alleviate the growing expense and inefficiency of citrus production. By supporting the commercial pomegranate venture with financial and technical resources, an external actor helped the Turkish Cypriot realize crop switching and overcome the monetary and informational barriers to climate adaptation identified by Seo and Mendelsohn (2008) and Huda et al. (2005).

We argue that the Turkish Cypriot farmers, like the Turkish Cypriot beekeepers studied by Gokcekus et al. (2014), were constrained by the classic collective action problem. The individual farmers did not have the resource capacity to transition to an alternative crop. They lacked the necessary training and technical background to implement a successful crop-switching programme, and they also faced financial constraints. Furthermore, their resource challenges made collectivization unlikely as both technical and financial leadership were absent. Once the collective action problem was overcome, we attribute the success of the pomegranate programme to its incorporation of sustainable peacebuilding (Donais 2009). An external actor provided technical knowledge and the initial investment funds, but Turkish Cypriot farmers quickly accepted local ownership of pomegranate farming. With external assistance, the domestic actor was able to solve the collective action problem.

4 THE EU AND USAID

Although the EU has committed to assisting north Cyprus' economic development in the hope of the eventual reunification of north and south Cyprus, it was not the primary external actor involved in the development of north Cyprus' new pomegranate industry. Managing Cyprus' political divisions has required delicate manoeuvring by the EU. Until 2013, the EU was actually "unable to set up a delegation in the Turkish-controlled half. Instead, it had to establish a headquarters-based task force in the south with a local programme support office in the northern part of Cyprus" (Nielsen 2012). With this in mind, the EU's relatively behind-the-scenes involvement in the pomegranate initiative is understandable. Through its Economic Development and Growth for Enterprises (EDGE) programme, USAID was actually the main actor facilitating the adoption of pomegranate farming. The EU's involvement was mostly restricted to providing funding through grants.

Under EDGE, the multinational consultancy firm BearingPoint was awarded a six million dollar contract to assist the TCC with economic development. In the completion of this contract, BearingPoint released certain online reports—including a 127 page final report—detailing the development of alternative crop programmes in north Cyprus. According to the EDGE reports, "the US Government's primary policy objective in Cyprus" was fostering a "durable settlement" between the divided Greek and Turkish Cypriot communities (EDGE2008b, p. 5). To facilitate this settlement, EDGE was commissioned "to improve banking practices, to strengthen business associations and services, and to provide firm-level assistance to promote enterprise competitiveness" (EDGE 2008a, p. 4). The programme was redefined to focus on sector-level initiatives after firm-level efforts failed to have a substantive impact on improving north Cyprus' competitiveness (EDGE2008a, p. 6). Within the agricultural sector, EDGE focused on five alternative crop programmes: pomegranates, capers, cactus fruit, salicornia, and passion fruit. The pomegranate programme, initiated in 2006, was the first of these to be implemented.

5 SELECTING POMEGRANATES

The decision to support the switch from citrus to pomegranate production was not so much related to the amount of water consumed by pomegranate trees as it was driven by the *type* of water they use. In fact, pomegranates trees require approximately the same amount of water as citrus trees (Sheets et al. 2013;

Sauls 1998; FAO 2013). What makes pomegranates a viable solution for Cyprus' climate limitations is their ability to "endure greater water salination than citrus trees" (EDGE 2008a, p. 16). Pomegranates trees are hardier than citrus and can better process salt-contaminated water. As an island in a frequent state of drought, Cyprus' freshwater resources are limited. Additionally, Cyprus has contaminated much of its available groundwater through over-extraction. As groundwater has been over pumped, freshwater aquifers have become salinated (Anastasi 2012, p. 14). EDGE reports specifically acknowledged the "salination of the ground water," as a contributing factor in Cyprus' declining citrus market competitiveness (EDGE 2008b, p. 87). Güzelyurt (its Greek name is Morphou), a region in north Cyprus, had been particularly affected by the increased salinity levels of the groundwater as it had significant citrus orchards. EDGE officials subsequently concentrated their initial pomegranate ventures in this region.

To further heighten the resource gains of the crop switch, the pomegranate varieties chosen by EDGE's commissioned experts were ones that flourished in drier climates and required fewer water resources. EDGE invited Agro Consultant and Agricultural Engineer Shlomy Raziel to provide expertise on the pomegranate programme. After analysing Cyprus' climate constraints, Raziel suggested that the phenology of the Wonderful variety might be best suited to the project as it had proved successful in similarly dry climates, such as Israel (EDGE 2008b, p. 66). Two other pomegranate varieties—the Herskovitz and the Acco that were both developed for Israel's climate—were also selected to be planted along with the Wonderful variety.

EDGE reports cited pomegranates' "consistent[cy] with the TCC's climate, topography and *marketing*[emphasis added] potential" as the reason they were selected to be the alternative crop initiative's prototype (EDGE 2008a, p. 33). As discussed, pomegranates had climate and environmental advantages over citrus fruits. However, the programme's ultimate success was also attributable to the market advantages of switching. Heightened competition from countries without the production difficulties of Cyprus' environmental constraints meant citrus producers were experiencing declining rates of return on their investments (EDGE2008b, p. 87). Simultaneously, demand for pomegranates increased due to the fruit's designation as a nutritional "superfruit" (Cassell 2012). Despite being a relatively new venture, market demand for the pomegranates produced through the EDGE programme has been such that participants have repeatedly sold the entirety of their yearly crop. All first quality fruits (those with traits most attractive to consumers) are consumed by the international market. The environmental gains of switching to pomegranate production were significant; the market benefits were equally important.

6 PROGRAMME DESIGN AND IMPLEMENTATION

Pomegranates are not a new fruit to Cyprus. During the nineteenth and early twentieth century, they were actually one of Cyprus' primary articles of export (Kahramanoglu et al. 2014; Usanmaz 2013, p. 2). The conflict that separated the island ended this trade and pomegranate orchards declined. Before the implementation of EDGE's alternative crops programme, pomegranates were either grown mostly for personal usage or "only remained around other fruit orchards as wind barriers" (Usanmaz et al. 2014, p. 62).

Prior to publicly advertising its alternative crop programmes, EDGE consultants exerted considerable effort in pre-emptively alleviating public backlash against perceived "replacement" crops (EDGE 2008b, p. 86). To protect the pomegranate programme from critics who claimed that it would subvert the livelihood of citrus growers, EDGE required that the land used by the growers be "unused, arid land" (p. 86). This stipulation meant that EDGE personnel had to devote additional resources to identifying land that fit this criteria and assisting the farmers with funding and developing drip irrigation systems to make the land productive. However, it did succeed at reducing local suspicions and had a positive long-term impact in the sense that in increased the amount of productive land available to Turkish Cypriot farmers. EDGE began publicly advertising its alternative crop programmes through various local channels in 2006. Of the five proposed projects, pomegranates received the greatest community interest. Twenty-two Turkish Cypriot farmers responded to the initial outreach efforts with 17 agreeing to EDGE's programme requirements. Farmers participating in the programme had to agree to the following:

(a) sign a Memorandum of Understanding (MoU) that outlined their responsibilities, (b) put some of their own monetary resources into the project, which usually came in the form of purchasing the plants, and (c) allow EDGE experts access to the alternative crops fields/orchards to inspect the crops and offer advice on how to achieve the best harvest. (EDGE 2008b, p. 86)

The programme's *Phase One* focused on planting the pomegranates. EDGE arranged for Shlomy Raziel to visit the pomegranate orchards every 6–8 weeks in order to provide the farmers with on-site technical expertise (EDGE 2008b, p. 86). Prior to the actual planting of the pomegranate trees, EDGE required the farmers to attend training sessions on financial management, orchard establishment, and tree planting. After the 17 participating farmers planted the 18,750 trees on 51 acres of formerly arid land, EDGE supplemented Mr. Raziel's assistance with additional sessions on fertilizer application, pest management, winter pruning, and flower thinning (EDGE 2008b, pp. 117–118). *Phase Two* of the programme expanded the number of participating farmers to 22, the planted acreage to 62 acres, and the number of trees to 22,450 (EDGE2008b, p. 87).

Phase Three and *Four* of the programme were designed to develop the business side of this agribusiness programme. Under *Phase Three*, EDGE worked with the farmers to establish a fruit processing facility, while*Phase Four* focused on ensuring the sustainability of the growers' business model. Within *Phase Three*, 19 of the farmers incorporated the limited liability company Alnar Narcılık Ltd, and then using an EU grant of 150,000 €, they built a pomegranate processing and storage centre. Through Alnar, the farmers have been able to export their products to five EU countries: Sweden, England, Belgium, Germany, and the Netherlands. The collective has plans to extend their international reach beyond these countries and to expand their internationally available product line to include pomegranate juice. Table 1 provides a summary timeline of the pomegranate programme's various components.

Table 1: Timeline of pomegranate production

Year	Step
2006	EDGE develops agribusiness programmes
2006	Presented the programme to interested growers
2006	Signed MOUs with the 17 participants in phase 1
2006–2008	Training for growers
2006–2007	Establishment of the orchards
2007	Planted pomegranate trees
2007	Installed irrigation systems in orchards
2008	First harvest
2008	Establishment of Alnar Ltd
2010	Alnar awarded EU rural development grant
2011	Establishment of the Alnar processing facility
2011	First international export of pomegranates

7 SUSTAINABLE PEACEBUILDING

The pomegranate programme has proven to be an archetype upon which to base future sustainable agribusiness development programmes. As the farmers have become more adept at growing pomegranates, the amount of fruit produced per tree has doubled from that of the initial yield (FreshFruit 2014). Exports have grown to almost five times Alnar's preliminary offering of 49 t;

in 2013 Alnar exported 232 t of fresh pomegranates (Alan 2013). The collective now produces around 30,000 L of pomegranate juice; since the producers do not use preservatives in their purely natural pomegranate juice, this is currently only consumed by the local market (FreshPlaza 2012). However, the company is exploring alternative means of bottling their juice in order to access new markets.

EDGE's pomegranate programme flourished due to elements of sustainable peacebuilding, like "negotiated hybridity" and "local ownership", that were built into the project. From the beginning, EDGE required the local actors to accept ownership of the programme through financial and organizational commitments. The monetary resources committed to the programme by the farmers were not insubstantial. Although USAID did provide some funding, its financial resources were mainly put towards training sessions and providing the farmers with technical assistance. As mentioned, the EU gave the Alnar collective a 150,000 € grant for the building of the processing plant. However, the entire project cost 450,000 €; after EDGE personnel assisted the farmers with developing a preliminary business plan, the local farmers were responsible for developing their own marketing strategy and contributing the remaining 300,000 € to the programme.

The local farmers were encouraged by the EDGE consultants to collectivize in order to foster long-term accountability among the farmers, develop local leadership, and ensure the sustainability of the programme. According to Ibrahim Kahramanoglu, managing director of Alnar, the producers collectivized in order to have ownership over the programme's business decisions. EDGE personnel suggested that Alnar apply for GLOBALGAP. certification, an independently evaluated standard designation that is often required by EU buyers. EDGE assisted Alnar with the initial application; recertifications have been handled by Alnar's personnel. After obtaining their initial certification, Alnar negotiated with the Cyprus Pomegranate Producers Union (a body of 36 pomegranate producers that has significant overlap with the producers of the Alnar collective) to handle exportation of their crops and extend Alnar's certification to cover the pomegranate products of the Union. Through Alnar, the pomegranate producers were able to develop the local leadership necessary to sustain and surpass what they achieved under the guidance of an external actor.

8 POLICY IMPLICATIONS

Cyprus is certainly not the only country facing climate-related development challenges nor is it the only developing country with a collective action problem. Thus, the strategies that worked to foster sustainable pomegranate farming should be largely applicable elsewhere and to other agricultural products. From the case of pomegranate farming, the importance of the following factors is apparent.

When implementing crop switching, the market reputations of suggested replacement products carry considerable weight in determining whether one product is adopted over another. Both EDGE and the local farmers cited pomegranate's rapidly expanding reputation as a superfruit as an important factor in its selection. Even with the programme's success, the director of Alnar complained that Europe lacked a "pomegranate culture" and that pomegranate products were still far from reaching optimal market recognition (FreshFruitPortal 2014). Although the Turkish Cypriot farmers did have an interest in pomegranates prior to external involvement, identifying products that are both environmentally and market friendly is potentially an area in which developing countries need assistance.

The case of north Cyprus also demonstrated that locals were more likely to view crop-switching programmes favourably if they were initially perceived as augmenting, rather than replacing, traditional crops. EDGE's approach of planting pomegranates only on unused, arid land was able to mitigate any complaints from citrus producers. However, this approach did require significant exertion on the part of the external actor to both identify the available land and then provide the technical resources to make this land arable.

Local awareness of the suggested alternative crop also impacted the programme's overall success. Of the five alternative crop programmes developed by EDGE—pomegranates, cactus, fruit, salicornia, capers, and passion fruit—the pomegranate programme has been most successful. Its success is at least partially attributable to the farmers' familiarity with pomegranates; the passion fruit programme has not inspired nearly as much interest among local actors. In interviews, the director of Alnar stressed that the pomegranate programme only succeeded (and was able to become highly successful) because of the awareness and enthusiasm of the local farmers for pomegranates. In addition to selecting crops that are both market and environmentally friendly, external actors also need to consider local preferences when designing development projects.

Local ownership is critical to achieving successful, sustainable peacebuilding through development. As is evident from the experience of the pomegranate producers, one of the simplest ways in which to inspire ownership is to require significant monetary commitments from the local participants. Overcoming the collective action problem that often characterizes developing societies does require financial input by the external actor (in order to pay for the necessary technical specialists); however, this input should be matched (within reason and considering circumstances) by local contributions. Equally important as the technical training it provided was the assistance EDGE gave the farmers in creating a business strategy. Motivated by EDGE, the producers were able to capitalize on their training and expand into exporting. Thus, in the case of Cyprus, the most important factor was not how much money was given, but how that money was used.

Fig. 1 A sign indicating that this is a GLOBALGAP certified pomegranate orchard, in Morphou area, Cyprus

9 CONCLUDING COMMENTS

Pomegranate farming in north Cyprus might appear to be a rather niche case. However, it highlights a real problem that both developing and developed countries will soon—if they have not already—confront, namely, climate change. As the evidence from Cyprus indicates, post-conflict developing countries need financial and technical intervention from an external actor in order to implement even the simplest of climate change adaptations, such as crop switching. Additionally, post-conflict states are often hobbled by the classic collective action problem and need an outside impetus to inspire them to accept local ownership of a development programme. Sustainable development is therefore a function of external and local actor coordination; using its superior resources, the external actor helps initiate the programme, and since the local actor also makes a significant contribution, the local actor commits to ensuring the programme's success.

The sustainability of similar climate change programmes can be further enhanced by selecting crops that are at least somewhat familiar to the local community and are also desired by the external market. A crop's phenology might be ideally suited to a certain region, but the local farmers need to also support its production. Identifying suitable crops will require the combined effort of external and local actors. Ultimately, enacting successful climate change adaptations requires both parties to share leadership responsibilities (Fig. 1).

FOOTNOTES

1 Any information not directly cited is attributable to interviews and personal correspondence with Berna Berberoğlu, the deputy project manager of the Economic Development and Growth for Enterprises

(EDGE) programme, and İbrahim Kahramanoğlu, the managing director of Alnar Narcılık Ltd.

REFERENCES

1. Alnar Narcılık Ltd (2013) Project—Alnar Narcılık Ltd Projesi, AB Bilgi Merkezi. http://vimeo.com/63579743

2. Alan D (2013) Dünyaya nar satıyoruz. Havadis, December 23, 2013. http://www.havadiskibris.com/Haberler/kibris/dunyaya-nar-satiyoruz/16757

3. Alnar Narcilik Ltd (2012) Mission. http://www.alnarpom.com/

4. Anastasi E (2012) The impact of climate change in water resources in Cyprus and possible human health effects. Cyprus University of Technology, Cyprus, pp 2–19. http://cyprus-institute.us/Student-Records/2012/2012-Student%20Records/Elena_Anastasi/elena%20anastasi_climate%20change%20paper.pdf

5. BBC (2010) Cyprus conflict closes leaders' eyes to water shortage. March 11, 2010. http://news.bbc.co.uk/2/hi/programmes/from_our_own_correspondent/8560424.stm

6. BBC (2011) Timeline: Cyprus. http://news.bbc.co.uk/2/hi/europe/country_profiles/1021835.stm

7. Cassell D (2012) Pomegranate retains superfruit status. Food Processing Magazine. http://www.foodprocessing.com/articles/2012/pomegranate-retains-superfruit-status/

8. Cyprus Profile (2013) Agriculture and Food: Quality not quantity. July 2013. http://www.cyprusprofile.com/en/sectors/agriculture-and-food

9. Donais T (2009) Empowerment or imposition? Dilemmas of local ownership in post-conflict peacebuilding processes. Peace Change 34:3–26

10. Dworak T, Berglund, M (2012) The role of water pricing and water allocation in agriculture in developing sustainable water use in Europe. European Commission Final Report Project Number 11589

11. EDGE (2008a) Unlocking the Turkish Cypriot Community's potential by providing keys to economic growth.http://pdf.usaid.gov/pdf_docs/pdacx162.pdf

12. EDGE (2008b) Final Report. Unlocking the Turkish Cypriot Community's potential by providing keys to economic growth, August 2008. http://pdf.usaid.gov/pdf_docs/pdacy313.pdf

13. EEA (2009) Water resources across Europe—confronting water scarcity and drought. EEA Report 2/2009

14. FAO (2013) Crop water information: citrus. http://www.fao.org/nr/water/cropinfo_citrus.html

15. FreshFruitPortal (2014) Pomegranate culture lacking in Europe, says Cypriot grower. February 5, 2014

16. FreshPlaza (2012) Turkey: ALNAR full range of pomegranate products. February 28, 2012. http://www.freshplaza.com/article/93455/Turkey-ALNAR-full-range-of-pomegrate-prodcts

17. Gies E (2013) Northern Cyprus sees hope in water pipeline. The New York Times, April 3, 2013. http://www.nytimes.com/2013/04/04/world/europe/northern-cyprus-sees-hope-in-water-pipeline.html?pagewanted=all&_r=0

18. Gokcekus O (2008) The economics of the isolation of Turkish Cypriots. Turkish Cypriot Chamber of Commerce, Freebirds, Kyrenia

19. Gokcekus O, Bengyak K (2014) Peculiar dynamics of corruption: religion, gender, EU membership, and others. World Scientific, Singapore

20. Gokcekus O, Finnegan C, Cakal H (2014) External actors' role in solving local collective action problems in a post-conflict setting: a case study of the Turkish Cypriot beekeepers. Econ Peace Secur J 2:19–26

21. Huda S, Sadras V, Wani S, Mei X (2005) Food security and climate change in the Asia-Pacific region: evaluating mismatch between crop development and water availability. Eur Foreign Aff Rev 10:535–552

22. Kahramanoglu I, Usanmaz S, Nizam I (2014) Incidence of heart rot at pomegranate fruits caused by Alternaria spp. in Cyprus. Afr J Agric Res 9:905–907

23. Kurukulasuriya P, Mendelsohn R (2008) Crop switching as a strategy for adapting to climate change. Afr J Agric Resour Econ 2:105–126

24. Lobell D, Burke M, Tebaldi C, Mastrandrea M, Falcon W, Naylor R (2008) Prioritizing climate change adaptation needs for foods security in 2030. Science 319:607–610

25. Nielsen N (2012) EU assistance to north Cyprus tangled in conflict. EUObserver, May 25, 2012

26. Olson M (1965) The logic of collective action. Harvard University Press, Cambridge, MA

27. Sands D (2014) Turkey builds massive pipeline to send drinking water to thirsty Cyprus. The Washington Times,

September 30, 2014. http://media.washtimes.com/media/misc/2014/10/01/northerncyprus_revisedv3.pdf

28. Sauls J (1998) Home fruit production-pomegranate. Texas Citrus and Subtropical Fruits, December 1998. http://aggie-horticulture.tamu.edu/citrus/pomegranate.htm

29. Seo N, Mendelsohn R (2008) An analysis of crop choice: adapting to climate change in South America farms. Ecol Econ 67:109–116

30. Sheets M, Du Bois M, Williamson J (2013) The pomegranate. University of Florida. http://edis.ifas.ufl.edu/mg056

31. Shoukri E, Zachariadis T (2012) Climate change in Cyprus: impacts and adaptation policies. Environmental Policy Research Group Report 01-12

32. Sofroniou A, Bishop S (2014) Water scarcity in Cyprus: a review and call for integrated policy. Water 6:2898–2928

33. Usanmaz S (2013) Current situation and future of pomegranate production in Turkish Republic of Northern Cyprus. Master of Science Thesis, Lefke

34. Usanmaz S, Kahramanoglu İ, Yılmaz N (2014) Yield and pomological characteristics of three pomegranate (Punica granatum L) cultivars: wonderful, Acco and Herskovitz. Am J Agric For 2:61–65

35. Zachariadis T (2010) Residential water scarcity in Cyprus: impact of climate change and policy options. Water 2:788–814

3
CHAPTER

Potent health effects of pomegranate

Aida Zarfeshany, Sedigheh Asgary,[1] and Shaghayegh Haghjoo Javanmard

Department of Engineering, University of Cambridge, Trumpington Street, CB2 1PZ Cambridge, United Kingdom

ABSTRACT

Accumulating data clearly claimed that *Punica granatum* L. (pomegranate) has several health benefits. Pomegranates can help prevent or treat various disease risk factors including high blood pressure, high cholesterol, oxidative stress, hyperglycemia, and inflammatory activities. It is demonstrated that certain components of pomegranate such as polyphenols have potential antioxidant, anti-inflammatory, and anticarcinogenic effects. The antioxidant potential of pomegranate juice is more than that of red wine and green tea, which is induced through ellagitannins and hydrosable tannins. Pomegranate juice can reduce macrophage oxidative stress, free radicals, and lipid peroxidation. Moreover, pomegranate fruit extract prevents cell growth and induces apoptosis, which can lead to its anticarcinogenic effects. In addition, promoter inhibition of some inflammatory markers and their production are blocked via ellagitannins. In this article, we highlight different studies on the therapeutic effects of pomegranate and their suggested mechanisms of actions.

Keywords: Antioxidant, inflammatory activities, pomegranate

INTRODUCTION

Punica granatum L. (Pomegranate) is a long-lived and drought-tolerant plant. Arid and semiarid zones are popular for growing pomegranate trees. They are widely cultivated in Iran, India, and the Mediterranean countries such as Turkey, Egypt, Tunisia, Spain, and Morocco.[1] However, pomegranate is categorized as a berry but it belongs to its own botanical family, *Punicaceae*. The only genus is *Punica*, with one predominant species called *P. granatum*.[2]

The trees can grow up to 30 feet in height. The leaves are opposite, narrow, oblong with 3-7 cm long and 2 cm broad. It has bright red, orange, or pink flowers, which are 3 cm in diameter with four to five petals. Edible fruit has a rounded hexagonal shape, with 5-12 cm in diameter and weighing 200 g. The thick skin surrounds around 600 arils, which encapsulates the seeds.[3]

CHEMICAL COMPOSITION

Seed

About 18% of dried and cleaned white seeds are oil. The oil is rich in punicic acid (65%), which is a triple conjugated 18-carbon fatty acid [Figure 1]. There are some phytoestrogen compounds in pomegranate seeds that have sex steroid hormones similar to those in humankind. The 17-alpha-estradiol is a mirror-image version of estrogen.[3]

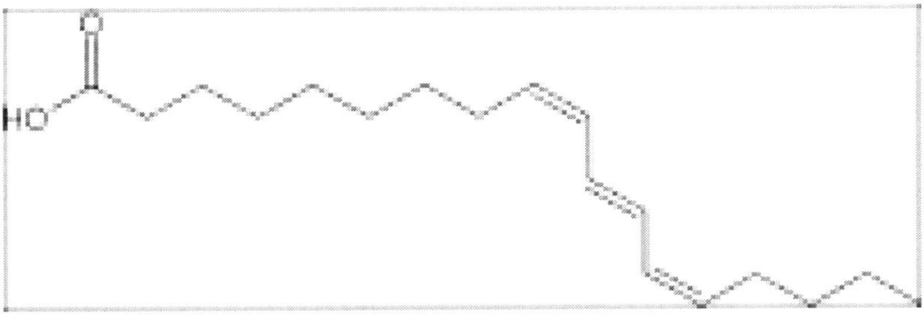

Figure 1 Chemical structure of punicic acid (9Z,11E,13Z-octadeca-9,11,13-trienoic acid)

Juice and peel

Pomegranate juice is a good source of fructose, sucrose, and glucose. It also has some of the simple organic acids such as ascorbic acid, citric acid, fumaric acid, and malic acid. In addition, it contains small amounts of all amino acids, specifically proline, methionine, and valine. Both the juice and peel are rich in polyphenols. The largest classes include tannins and flavonoids [Figure 2] that indicate pharmacological potential of pomegranate due to their strange antioxidative and preservative activities.[3]

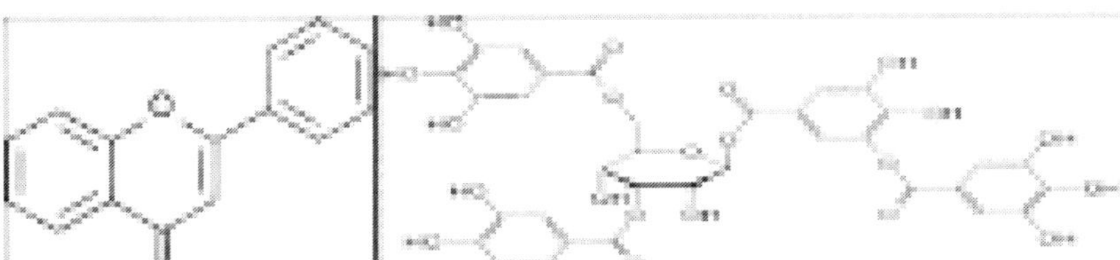

Figure 2 (a) Flavone backbone (2-phenyl-1, 4-benzopyrone). (b) Tannic acid

Ellagitannin is a type of tannins; it can be broken down into hydroxybenzoic acid such as ellagic acid. It is widely used in plastic surgeries, which prevents skin flap's death due to its antioxidant activity. Two other ellagitannins that are found in both pomegranate juice and peel are punicalagin and punicalin. Several classes of pomegranate flavonoids include anthocyanins, flavan 3-ols, and flavonols. Pomegranate juice and peel have catechins with a high antioxidant activity. They are essential compounds of anthocyanin's production with antioxidant and inflammatory role. Anthocyanins cause the red color of juice, which is not found in the peel. All pomegranate flavonoids show antioxidant activity with indirect inhibition of inflammatory markers such as tumor necrosis factor-alpha (TNF-α).[3]

Bark and roots

The pomegranate tree's bark and roots are rich sources of chemicals called alkaloids. They are carbon-based substances; they were used to treat worms in the human gastrointestinal tract in traditional medicine.[3]

Table 1 Shows pomegranate's nutrient values for 100 g of raw edible portion[4]

Table 1 Pomegranate's nutrient values for 100 g of raw edible portion

Table 1

Pomegranate's nutrient values for 100 g of raw edible portion

Nutrients	Units	Value per 100 g
Water	g	77.93
Energy	Kcal	83
Protein	g	1.67
Total lipid (fat)	g	1.17
Ash	g	0.53
Carbohydrates	g	18.70
Fiber	g	4.0
Sugars, total	g	13.67
Calcium	Mg	10
Iron	Mg	0.3

Health effects

Prostate cancer

After lung cancer, the second leading cause of male cancer death is prostate cancer worldwide. Its progress before onset of symptoms is slow; therefore, pharmacological and nutritional interventions could affect the quality of patient's life by delaying its development.[5]

It was shown that pomegranate fruit could be used in the treatment of human prostate cancer because it could inhibit cell growth and induce apoptosis.[6] It leads to induction of pro-apoptotic proteins (Bax and Bak) and downregulation of anti-apoptotic proteins (Bcl-xL and Bcl-2).[6] Moreover, the presence of NFκB and cell viability of prostate cancer cell lines has been inhibited when using pomegranate fruit extract, because it blocks NFκB.[7] Polyphenols of fermented juice and pomegranate oil can inhibit the proliferation of LNCaP (epithelial cell line derived from a human prostate carcinoma), PC-3, and DU145 human prostate cancer cell lines. These effects were the result of changes in cell cycle distribution and apoptosis induction.[6] In addition, it is reported that pomegranate fruit extract oral administration in nude mice implanted with androgen-sensitive CWR22RV1 cells caused significant decrease in serum prostate-specific antigen (PSA) level and inhibited tumor growth.[7] Besides, the observed increase in NFκB activity during androgen dependence to androgen independence transition in the LAPC4 xenograft model was terminated.[8]

Breast cancer

Fermented pomegranate juice has double the antiproliferative effect compared to fresh pomegranate juice in human breast cancer cell lines MCF-7 (breast cancer cell line isolated in 1970 from a 69-year-old Caucasian woman) and MB-MDA-231. In addition, pomegranate seed oil caused 90% prevention of proliferation of MCF-7 cells.[9,10]

Lung cancer

Pomegranate fruit extract can inhibit several signaling pathways, which can be used in the treatment of human lung cancer. Pathways include Mitogen-activated protein kinases (MAPK) PI3K/Akt and NFκB. In addition, there was a 4 day delay in the appearance of tumors (from 15 to 19 days) in mice implanted with A549 cells.[10] These studies indicate the chemopreventive effects of pomegranate fruit extract.[3]

Colon cancer

Adams *et al.*[11] have reported the anti-inflammatory effects of pomegranate juice on the signaling proteins in HT-29 human colon cancer cell line. Reduction in phosphorylation of the p65 subunit of NFκB, its binding to the NFκB response, and 79% inhibition in TNF-α protein expression have been observed with 50 mg/L concentration of pomegranate extract.

Skin cancer

It has been demonstrated that pomegranate oil has chemopreventive efficacy in mice. Reduced tumor incidence (7%), decrease in tumor numbers, reduction in ornithine decarboxylase (ODC) activity (17%), significant inhibition in elevated Tissue plasminogen activator (TPA)-mediated skin edema and hyperplasia, protein expression of ODC and COX-2, and epidermal ODC activity have been reported with pomegranate oil treatments.[12,13] Pomegranate extract in various concentrations (5-60 mg/L) was effective against UVA- and UVB-induced damage in SKU-1064 fibroblast cells of human, which was relevant in reducing NFκB transcription, downregulating proapoptotic caspase-3, and elevating the G0/G1 phase associated with deoxyribonucleic acid (DNA) repair.[14]

Cardiovascular diseases

Pomegranate juice is an affluent source of polyphenols with high antioxidative potential. Moreover, its antiatherogenic, antihypertensive, and anti-inflammatory effects have been shown in limited studies in human and murine models.[15]

Hypertension is the most common disease in primary care of patients. It is found in comorbidity with diabetes and cardiovascular disease, and the majority of patients do not tend to be medicated.

Pomegranate juice prevents the activity of serum angiotensin-converting enzyme and reduces systolic blood pressure.[16] Angiotensin II acute subcutaneous administration causes increased blood pressure in diabetic Wistar rats. It has been shown that pomegranate juice administration (100 mg/kg) for 4 weeks could reduce the mean arterial blood pressure.[17] Pomegranate juice consumption resulted in 30% decrease in carotid intima-media thickness after 1 year. The patient's serum paraoxonase 1 (PON 1) activity showed 83% increase, whereas both serum low dwnsity lipoprotein (LDL) basal oxidative state and LDL susceptibility to copper ion significantly decreased by 90% and 95%, respectively.[18]

Punicic acid, which is the main constituent of pomegranate seed oil, has antiatherogenic effects. In a study on 51 hyperlipidemic patients, pomegranate seed oil was administered twice a day (800 mg/day) for 4 weeks. There was a significant decrease in triglycerides (TG) and TG: High density lipoprotein (HDL) cholesterol ratio by 2.75 mmol/L and 5.7 mmol/L, respectively, whereas serum cholesterol, LDL-C, and glucose concentration remained unchanged.[19]

High plasma LDL concentration is the major risk factor for atherosclerosis. Therefore, LDL modifications, including oxidation, retention, and aggregation, play a key role in atherosclerosis as well. Studies have shown that consuming pomegranate juice for 2 weeks resulted in declined retention and aggregation of LDL susceptibility and increased activity of serum paraoxonase (a protective lipid peroxidation esterase related to HDL) by 20% in humans. Pomegranate juice administration in mice for 14 weeks showed reduced LDL oxidation by peritoneal macrophages by more than 90%, which was because of reduced cellular lipid peroxidation and superoxide release. The uptake of oxidized LDL showed 20% reduction in mice. The size of atherosclerotic lesions reduced by 44% after pomegranate juice supplementation.[20] Moreover, pomegranate juice administration to apolipoprotein E-deficient mice with advanced atherosclerosis for 2 months reduced oxidized LDL (31%) and increased macrophage cholesterol efflux (39%).[21]

In cultured human endothelial cells and hypercholesterolemic mice, both pomegranate juice and fruit extract reduced the activation of ELK-1 and p-CREB (oxidation-sensitive responsive genes) and elevated the expression of endothelial nitric oxide synthase. It is suggested that polyphenolic antioxidant compounds in pomegranate juice are responsible for the reduction of oxidative stress and atherogenesis.[22]

In another study,[23] concentrated pomegranate juice was shown to reduce heart disease risk factors. Administration of concentrated pomegranate juice to 22 diabetic type 2 patients with hyperlipidemia could significantly reduce TC, LDL-C, LDL-C: HDL-C ratio, and TC: HDL-C ratio. However, it was unable to decrease serum TG and HDL-C concentrations.

Oral administration of pomegranate flower aqueous extract in streptozotocin (STZ)-induced albino Wistar rats in both 250 mg/kg and 500 mg/kg doses for 21 days could significantly reduce fibrinogen (FBG), TC, TG, LDL-C, and tissue lipid peroxidation level and increased the level of HDL-C and glutathione content.[24]

Heart fibrosis increases among diabetics, which results in impairing cardiac function. Endothelin (ET)-1 and NFκB are interactive fibroblast growth regulators. It is suggested that pomegranate flower extract (500 mg/kg/day) in Zucker diabetic fatty rats could reduce the ratios of van Gieson-stained interstitial collagen deposit area to a total left ventricular area and perivascular collagen deposit areas to coronary artery media area in the heart and diminishes cardiac fibrosis in these rats. In addition, overexpressed cardiac fibronectin and collagen I and II messenger RNAs (mRNAs) were inhibited. It also decreased the upregulated cardiac mRNA expression of ET-1, ETA, inhibitor-κBβ, and c-jun. Pomegranate flower extract is a dual activator of peroxisome proliferator-activated receptor (PPAR)-α and γ and improves hyperlipidemia, hyperglycemia, and fatty heart in diabetic fatty Zucker rats.[25,26]

Punicic acid caused a dose-dependent increase in PPAR alpha and gamma reporter activity in 3T3-L1 cells. Dietary punicic acid reduced plasma glucose, suppressed NFκB activation and unregulated TNF-α expression and PPAR-α/γ responsive genes in adipose tissue and skeletal muscle.[27]

Pomegranate leaf extract was administered (400 and 800 mg/kg/day) to high-fat-diet-induced obese and hyperlipidemic mouse models for 5 weeks. The results indicated significant reduction in body weight, energy intake (based on food intake), serum total cholesterol (TC), TG, FBG, and TC/HDL-C ratio. Intestinal fat absorption was inhibited as well.[28]

The high fat diet (HFD) with 1% pomegranate seed oil (rich source of punicic acid) was administered for 12 weeks to induce obesity and insulin resistance in mice. The pomegranate seed oil-fed group exhibited lower body weight (4%) and body fat mass (3.1%) compared with only HFD-fed mice. A clear improvement was observed in peripheral insulin sensitivity (70%) in pomegranate seed oil-administered rats.[29]

Fatty liver is the most common abnormal liver function among diabetics. Pomegranate flower was examined for its antidiabetic effects on diabetic type II and obese Zucker rats. Rats fed with 500 mg/kg/day of pomegranate flower extract for 6 weeks showed decreased ratio of liver weight to tibia length, lipid droplets, and hepatic TG contents. In addition, it increased PPRA-α and Acyl-COA oxidase mRNA levels in HepG2 cells.[30]

In a study by de Nigris *et al.*,[31] they compared the influence of pomegranate fruit extract with pomegranate juice on nitric oxide and arterial function in obese Zucker rats. They have demonstrated that both pomegranate fruit extract and juice significantly reduced the vascular inflammatory markers expression, thrombospondin, and cytokine TGFP 1. Increased plasma nitrite and nitrate were observed with administration of either pomegranate fruit or juice.

Many studies have reported the anti-inflammatory potential of pomegranate extract. In a study on 30 Sprague-Dawley rats with acute inflammation due to myringotomy, it was observed that 100 μl/day of pomegranate extract could significantly reduce reactive-oxygen species (ROS) levels. The extract was administered 1 day before and 2 days after surgery. Reduced thickness of lamina propria and vessel density was reported as well.[32] Both ellagitannins and ellagic acid are the main components of pomegranate extract, which have anti-inflammatory properties. They are metabolized by gut microbiota to yield urolithins. It is suggested that urolithins are the main components responsible for the anti-inflammation properties of pomegranate. It is suggested that NFκB activation, MAPK downregulation of COX-2, and mPGES-1 expression were inhibited through a decrease in PGE2 production.[33] Neutrophils play key roles in inflammatory processes by releasing great amounts of ROS generated by NADPH-oxidase and myeloperoxidase. It is indicated that punicic acid exhibited a potent anti-inflammatory effect via prevention of TNF-α-induced priming of NADPH oxidase by targeting the p38MAPKinase/Ser 345-p 47 phox-axis and releasing MPO.[34] Hyperglycemia results in oxidative stress in diabetes mellitus, which is a major factor in the pathogenesis of cardiovascular disease. Results suggested that pomegranate extract, owing to its polyphenol-rich antioxidants (oleanolic, ursolic, and gallic acids), could prevent cardiovascular complications through decrease in LDL, increase in HDL, serum paraoxonase 1 stability and activity, and nitric oxide production.[35,36,37]

Osteoarthritis

The most common forms of arthritis are osteoarthritis and its major progressive degenerative joint disease, which could affect joint functions and quality of life in patients. It is mediated by proinflammatory cytokines such as IL-1 and TNF-α. MAPKs are important due to their inflammatory and cartilage damage regulation.[38] P38-MAPKs are responsible for regulating cytokine production, neutrophils activation, apoptosis, and nitric oxide synthesis. The MAPK family phosphorylates a number of transcription factors such as runt-related transcription factor-2 (RUNX-2).[39,40,41]

Pomegranate extract, with its rich source of polyphenols, can inhibit IL-1 β-induced activation of MKK3, DNA-binding activity of RUNX-2 transcription factor, and p38 α-MAPK isoform.[38]

Rheumatoid arthritis

Rheumatoid arthritis is an autoimmune disease that affects 0.5-1% of people worldwide. Women are afflicted more than men. This inflammatory disease is characterized by inflammation and bone erosion.[38,39] Critical mediators in the pathogenesis of rheumatoid arthritis are TNF-α, IL-1 β, MCP1, Inducible nitric oxide synthase (iNOS), and COX-2-agents, which are stimulated by p38-MAPK and NFκB activation.[42,43]

It is shown that pomegranate extract could reduce the onset and incidence of collagen-induced arthritis in mice. Severity of arthritis, joint inflammation, and IL-6 level were significantly reduced in pomegranate extract-fed mice.[44]

Antimicrobial/fungal effect

Since bacterial resistance to antimicrobial drugs is increasing, medicinal plants have been considered as alternative agents. Pomegranate has been widely approved for its antimicrobial properties.[4,45,46] It has been shown that dried powder of pomegranate peel has a high inhibition of*Candida albicans*.[47] In addition, antimicrobial effects of both methanol and dichloromethane pomegranate extracts have been demonstrated on the *Candida* genus yeast as pathogen-causing disease in immunosuppressive host.[48] Methicillin-resistant *staphylococcus aureus* (MRSA) and methicillin-sensitive*staphylococcus aureus* (MSSA) (multiple antibiotics resistant) produce panta valentine leukocidin (PVL) toxin, which can lead to higher levels of morbidity and mortality.[49,50] It is indicated that a combination of pomegranate peel extract with Cu (II) ions exhibit enhanced antimicrobial effects against isolated MSSA, MRSA, and PVL.[51] One of the leading etiological bacteria of urinary tract infections is *Escherichia Coli*. Strong antibacterial activity of ethanol extract against *E. coli* has been shown.[52]

Skin

Solar ultraviolet radiations are the primary causes of many biological effects such as photoaging and skin cancer. These radiations resulted in DNA damage, protein oxidation, and matrix metalloproteinases induction. In one study, the effects of pomegranate juice, extract, and oil were examined against UVB-mediated damage. These products caused a decrease in UVB-induced protein expression of c-Fos and phosphorylation of c-Jun.[53] On the other hand, production of proinflammatory cytokines IL-1 β and IL-6 was decreased by topical application of 10 micromol/L of ellagic acid. The inflammatory macrophages infiltration was blocked in the integuments of SKH-1 hairless UVB-exposed mice for 8 weeks.[54]

Dental effects

The interbacterial coaggregations and these bacterial interactions with yeasts are related to the maintenance of oral microbiota. It is indicated that dried, powdered pomegranate peel shows a strong inhibition of *C. albicans* with a mean zone of 22 mm.[55] In another study, the antiplaque effect of pomegranate mouth rinse has been reported.[56] In addition, hydroalcoholic extract of pomegranate was very effective against dental plaque microorganisms (84% decrease (cfu/ml)).[57]

Reproductive system

One of the main constituents (16%) of the methanolic pomegranate seed extract is beta-sitosterol. It is suggested that the extract is a potent phasic activity stimulator in rat uterus, which happens due to the non-estrogenic effects of beta-sitosterol on inhibiting sarco-endoplasmic reticulum Ca^{2+} -ATPase (SERCA) and K channel, which resulted in contraction by calcium entry on L-type calcium channels and myosin light chain kinase (MLCK).[58] It is demonstrated that pomegranate fruit extract has an embryonic protective nature against adrianycin-induced oxidative stress (adrianycin is a chemotherapeutic drug used in cancer treatment).[59] Moreover, pomegranate juice consumption could increase epididymal sperm concentration, motility, spermatogenic cell density, diameter of seminiferous tubules and germinal cell layer thickness.[60]

Alzheimer

Hartman *et al.*[61] showed that mice treated by pomegranate juice have 50% less soluble Abeta 42 accumulation and amyloid deposition in the hippocampus, which could be considered for Alzheimer's disease improvement.

Malaria

In the presence of pomegranate fruit rind, the induced MMP-9 mRNA levels by haemozoin or TNF was decreased, which may be attributed to the antiparasitic activity and the inhibition of the proinflammatory mechanisms responsible in the onset of cerebral malaria.[62,63]

HIV

The anti-HIV-1 microbicide of pomegranate juice blocks virus binding to CD4 and CXCR4/CCR5, thereby preventing infection by primary virus clades A to G and group O.[64]

Wound healing

Use of pomegranate extract and flower showed significant reduction in wound area and increased the well-organized bands of collagen, fibroblasts, and few inflammatory cells.[65,66] Properties of elevated wound contraction and the period of epithelialization, collagen, and protein synthesis were reported in hydroalcoholic pomegranate extract.[67]

Mechanisms of action

Pomegranate can induce its beneficial effects through its various metabolites. The antioxidant and antiatherosclerotic potentials of pomegranate are mainly relevant to the high polyphenol concentrations in pomegranate fruit such as ellagitannins and hydrolysable tannins.[68] COX-1 and COX-2 enzymes and IL-1 β activity can be inhibited by pomegranate fruit extract.[69]

It is suggested that pomegranate can antagonize the stimulation of mRNA of MMP-9 in THP-1/monocytes. The whole fruit and compounds inhibit TNF-induced MMP-9 promoter activity.[41] Urolithins are metabolites that are metabolized by the human intestinal microflora. These compounds decreased MMP-9 sretion and mRNA levels induced by HZ or TNF. It is suggested that ellagitannins are responsible for the control of excessive production of MMP-9, which could result in decreased production of noxious cytokine TNF.[70] TNF cytokines promote NFκB binding to target sequences while inducing transcription of several genes such as the MMP-9 gene.[71] Ellagitannins prevent NFκB promoter activity by blocking NFκB-driven transcription and affecting the entire cytokine cascade. Ellagitannins inhibit the activation of inflammatory pathways such as MAPK.[40] In addition, pomegranate compounds could inhibit angiogenesis through the downregulation of vascular endothelial growth factor in cancers.[4]

Drug interactions involving pomegranate

Therapeutic benefits of pomegranate in various diseases would lead to an increase in its consumption.[72] It is important that pomegranate consumption does not affect the oral bioavailability of drugs.[73]

A study on human liver microsomes has shown the inhibitory effect of pomegranate juice on CYP2CP (a gene that codes for an enzyme to break down warfarin in the body) and increased bioavailability of tolbutamide (substrate for CYP2CP) in rats. Moreover, it is suggested that pomegranate may inhibit cytochrome P450-3A (CYP3A)-mediated carbamazepine metabolism.[4,74,75]

Pomegranate safety

Many studies have been carried out on the different components derived from pomegranate but no adverse effects have been reported in the examined dosage. Histopathological studies on both sexes of OF-1 mice confirmed the non-toxic effects of the polyphenol antioxidant punicalagin. Besides, in a study on 86 overweight human subjects who received 1420 mg/day of pomegranate fruit extract in tablet form for 28 days, no side effects or adverse changes in urine or blood of individuals were reported.[4,76]

Products and supplementation

Apart from fruit, pomegranate is available in various forms such as bottled juice (fresh or concentrated), powdered capsules, and tablets, which are derived from seed, fermented juice, peel, leaf and flower, gelatin capsules of seed oil extracts, dry or beverage tea from leaves or seeds, and other food productions such as jams, jellies, sauces, salad dressings, and vinegars. Anardana, which is the powdered form of pomegranate seed, is used as a form of spice.[3]

CONCLUSION

Pomegranate is a potent antioxidant. This fruit is rich in flavonoids, anthocyanins, punicic acid, ellagitannins, alkaloids, fructose, sucrose, glucose, simple organic acids, and other components and has antiatherogenic, antihypertensive, and anti-inflammatory properties. Pomegranate can be used in the prevention and treatment of several types of cancer, cardiovascular disease, osteoarthritis, rheumatoid arthritis, and other diseases. In addition, it improves wound healing and is beneficial to the reproductive system. Pomegranate can induce its beneficial effects through the influence of its various bioavailable constituents and metabolites on gene expression.

Although many *in vitro*, animal and clinical trials have been carried out to examine and prove the therapeutic effects of these compounds, further human trials and studies are necessary to understand the therapeutic potentials of pomegranate.

REFERENCES

1. Ercisli S, Gadze J, Agar G, Yildirim N, Hizarci Y. Genetic relationships among wild pomegranate (*Punica granatum*) genotypes from Coruh Valley in Turkey. Genet Mol Res. 2011;10:459–64.

2. Newman R. Sydney, Australia: Readhowyouwant; 2011. A wealth of phtochemicals. Pomegranate: The Most Medicinal Fruit (Large Print 16pt) p. 184.

3. Newman RA, Lansky EP, Block ML. Pomegranate: The Most Medicinal Fruit. Laguna Beach, California: Basic Health Publications; 2007. -A Wealth of Phytochemicals; p. 120.

4. Jurenka JS. Therapeutic applications of pomegranate (*Punica granatum* L.): A review. [Accessed September 2010];Altern Med Rev. 2008 13:128–44. USDA 2010. Pomegranates, Raw. United States Department of Agriculture.

5. Malik A, Afaq F, Sarfaraz S, Adhami VM, Syed DN, Mukhtar H. Pomegranate fruit juice for chemoprevention and chemotherapy of prostate cancer. Proc Natl Acad Sci U S A. 2005;102:14813–8.

6. Rettig MB, Heber D, An J, Seeram NP, Rao JY, Liu H, et al. Pomegranate extract inhibits androgen-independent prostate cancer growth through a nuclear factor-kappaB-dependent mechanism. Mol Cancer Ther. 2008;7:2662–71.

7. Albrecht M, Jiang W, Kumi-Diaka J, Lansky EP, Gommersall LM, Patel A, et al. Pomegranate extracts potently suppress proliferation, xenograft growth, and invasion of human prostate cancer cells. J Med Food.2004;7:274–83.

8. Kim ND, Mehta R, Yu W, Neeman I, Livney T, Amichay A, et al. Chemopreventive and adjuvant therapeutic potential of pomegranate (*Punica granatum*) for human breast cancer. Breast Cancer Res Treat.2002;71:203–17.

9. Mehta R, Lansky EP. Breast cancer chemopreventive properties of pomegranate (*Punica granatum*) fruit extracts in a mouse mammary organ culture. Eur J Cancer Prev. 2004;13:345–8.

10. Khan N, Hadi N, Afaq F, Syed DN, Kweon MH, Mukhtar H. Pomegranate fruit extract inhibits prosurvival pathways in human A549 lung carcinoma cells and tumor growth in athymic nude mice.Carcinogenesis. 2007;28:163–73.

11. Adams LS, Seeram NP, Aggarwal BB, Takada Y, Sand D, Heber D. Pomegranate juice, total pomegranate ellagitannins, and punicalagin suppress inflammatory cell signaling in colon cancer cells. J Agric Food Chem. 2006;54:980–5.

12. Hora JJ, Maydew ER, Lansky EP, Dwivedi C. Chemopreventive effects of pomegranate seed oil on skin tumor development in CD1 mice. J Med Food. 2003;6:157–61.

13. Syed DN, Malik A, Hadi N, Sarfaraz S, Afaq F, Mukhtar H. Photochemopreventive effect of pomegranate fruit extract on UVA-mediated activation of cellular pathways in normal human epidermal keratinocytes. Photochem Photobiol. 2006;82:398–405.

14. Pacheco-Palencia LA, Noratto G, Hingorani L, Talcott ST, Mertens-Talcott SU. Protective effects of standardized pomegranate (*Punica granatum* L.) polyphenolic extract in

ultraviolet-irradiated human skin fibroblasts. J Agric Food Chem. 2008;56:8434–41.

15. Gil MI, Tomás-Barberán FA, Hess-Pierce B, Holcroft DM, Kader AA. Antioxidant activity of pomegranate juice and its relationship with phenolic composition and processing. J Agric Food Chem.2000;48:4581–9.

16. Stowe CB. The effects of pomegranate juice consumption on blood pressure and cardiovascular health.Complement Ther Clin Pract. 2011;17:113–5.

17. Mohan M, Waghulde H, Kasture S. Effect of pomegranate juice on Angiotensin II-induced hypertension in diabetic Wistar rats. Phytother Res. 2010;24:S196–203.

18. Aviram M, Rosenblat M, Gaitini D, Nitecki S, Hoffman A, Dornfeld L, et al. Pomegranate juice consumption for 3 years by patients with carotid artery stenosis reduces common carotid intima-media thickness, blood pressure and LDL oxidation. Clin Nutr. 2004;23:423–33.

19. Mirmiran P, Fazeli MR, Asghari G, Shafiee A, Azizi F. Effect of pomegranate seed oil on hyperlipidaemic subjects: A double-blind placebo-controlled clinical trial. Br J Nutr. 2010;104:402–6.

20. Aviram M, Dornfeld L, Rosenblat M, Volkova N, Kaplan M, Coleman R, et al. Pomegranate juice consumption reduces oxidative stress, atherogenic modifications to LDL, and platelet aggregation: Studies in humans and in atherosclerotic apolipoprotein E-deficient mice. Am J Clin Nutr. 2000;71:1062–76.

21. Kaplan M, Hayek T, Raz A, Coleman R, Dornfeld L, Vaya J, et al. Pomegranate juice supplementation to atherosclerotic mice reduces macrophage lipid peroxidation, cellular

cholesterol accumulation and development of atherosclerosis. J Nutr. 2001;131:2082–9.

22. de Nigris F, Williams-Ignarro S, Sica V, Lerman LO, D'Armiento FP, Byrns RE, et al. Effects of a pomegranate fruit extract rich in punicalagin on oxidation-sensitive genes and eNOS activity at sites of perturbed shear stress and atherogenesis. Cardiovasc Res. 2007;73:414–23.

23. Esmaillzadeh A, Tahbaz F, Gaieni I, Alavi-Majd H, Azadbakht L. Cholesterol-lowering effect of concentrated pomegranate juice consumption in type II diabetic patients with hyperlipidemia. Int J Vitam Nutr Res. 2006;76:147–51.

24. Bagri P, Ali M, Aeri V, Bhowmik M, Sultana S. Antidiabetic effect of *Punica granatum* flowers: Effect on hyperlipidemia, pancreatic cells lipid peroxidation and antioxidant enzymes in experimental diabetes. Food Chem Toxicol. 2009;47:50–4.

25. Huang TH, Yang Q, Harada M, Li GQ, Yamahara J, Roufogalis BD, et al. Pomegranate flower extract diminishes cardiac fibrosis in Zucker diabetic fatty rats: Modulation of cardiac endothelin-1 and nuclear factor-kappaB pathways. J Cardiovasc Pharmacol. 2005;46:856–62.

26. Huang TH, Peng G, Kota BP, Li GQ, Yamahara J, Roufogalis BD, et al. Pomegranate flower improves cardiac lipid metabolism in a diabetic rat model: Role of lowering circulating lipids. Br J Pharmacol.2005;145:767–74.

27. Hontecillas R, O'Shea M, Einerhand A, Diguardo M, Bassaganya-Riera J. Activation of PPAR gamma and alpha by punicic acid ameliorates glucose tolerance and suppresses obesity-related inflammation. J Am Coll Nutr. 2009;28:184–95.

28. Lei F, Zhang XN, Wang W, Xing DM, Xie WD, Su H, DU LJ. Evidence of anti-obesity effects of the pomegranate leaf

extract in high-fat diet induced obese mice. Int J Obes (Lond) 2007;31:1023–9.

29. Vroegrijk IO, van Diepen JA, van den Berg S, Westbroek I, Keizer H, Gambelli L. Pomegranate Seed Oil, a rich source of Punicic Acid, prevents diet-induced obesity and insulin resistance in mice. Food Chem Toxicol. 2011;49:1426–30.

30. Xu KZ, Zhu C, Kim MS, Yamahara J, Li Y. Pomegranate flower ameliorates fatty liver in an animal model of type 2 diabetes and obesity. J Ethnopharmacol. 2009;123:280–7.

31. de Nigris F, Balestrieri ML, Williams-Ignarro S, D'Armiento FP, Fiorito C, Ignarro LJ, et al. The influence of pomegranate fruit extract in comparison to regular pomegranate juice and seed oil on nitric oxide and arterial function in obese Zucker rats. Nitric Oxide. 2007;17:50–4.

32. Kahya V, Meric A, Yazici M, Yuksel M, Midi A, Gedikli O. Antioxidant effect of pomegranate extract in reducing acute inflammation due to myringotomy. J Laryngol Otol. 2011;1:370–5.

33. González-Sarrías A, Larrosa M, Tomás-Barberán FA, Dolara P, Espín JC. NF-kappaB-dependent anti-inflammatory activity of urolithins, gut microbiota ellagic acid-derived metabolites, in human colonic fibroblasts. Br J Nutr. 2010;104:503–12.

34. Boussetta T, Raad H, Lettéron P, Gougerot-Pocidalo MA, Marie JC, Driss F, et al. Punicic acid a conjugated linolenic acid inhibits TNFalpha-induced neutrophil hyperactivation and protects from experimental colon inflammation in rats. PLoS One. 2009;4:e6458.

35. Katz SR, Newman RA, Lansky EP. *Punica granatum*: Heuristic treatment for diabetes mellitus. J Med Food. 2007;10:213–7.

36. Rock W, Rosenblat M, Miller-Lotan R, Levy AP, Elias M, Aviram M. Consumption of wonderful variety pomegranate juice and extract by diabetic patients increases paraoxonase 1 association with high-density lipoprotein and stimulates its catalytic activities. J Agric Food Chem. 2008;56:8704–13.

37. Fenercioglu AK, Saler T, Genc E, Sabuncu H, Altuntas Y. The effects of polyphenol-containing antioxidants on oxidative stress and lipid peroxidation in Type 2 diabetes mellitus without complications. J Endocrinol Invest. 2010;33:118–24.

38. Rasheed Z, Akhtar N, Haqqi TM. Pomegranate extract inhibits the interleukin-1b-induced activation of MKK-3, p38a-MAPK and transcription factor RUNX-2 in human osteoarthritis chondrocytes -Arthritis Res Ther. 2010;12:195.

39. Lee JC, Laydon JT, McDonnell PC, Gallagher TF, Kumar S, Green D. A protein kinase involved in the regulation of inflammatory cytokine biosynthesis. Nature. 1994;372:739–46.

40. Kumar S, Votta BJ, Rieman DJ, Badger AM, Gowen M, Lee JC. IL-1- and TNF-induced bone resorption is mediated by p38 mitogen activated protein kinase. J Cell Physiol. 2001;187:294–303.

41. Loeser RF, Erickson EA, Long DL. Mitogen-activated protein kinases as therapeutic targets in osteoarthritis. Curr Opin Rheumatol. 2008;20:581–6.

42. Hayden MS, Ghosh S. Signaling to NF-B. Genes Dev. 2004;18:2195.

43. Schieven GL. The biology of p38 kinase: A central role in inflammation. Curr Top Med Chem.2005;5:921–8.

44. Shukla M, Gupta K, Rasheed Z, Khan KA, Haqqi TM. Bioavailable constituents/metabolites of pomegranate

(*Punica granatum* L) preferentially inhibit COX2 activity *ex vivo* and IL-1beta-induced PGE2 production in human chondrocytes *in vitro*. J Inflamm (Lond) 2008;5:9.

45. Lansky E, Shubert S, Neeman I. Pharmacological and therapeutic properties of pomegranate. Israel: CIHEAM-Options Mediterraneennes; 2004;42:231–5.

46. Satish S, Mohana D, Ranhavendra M, Raveesha K. Antifungal activity of some plant extracts against important seed borne pathogens of Aspergillus sp. J Agric Sci Technol. 2007;3:109–19.

47. Mithun P, Prashant G, Murlikrishna K, Shivakumar K, Chandu G. Antifungal efficacy of *Punica granatum*, *Acacia nilotica*, *Cuminum cyminum* and *Foeniculum vulgare* on Candida albicans: An *in vitro*study. Indian J Dent Res. 2010;21:334–6.

48. Höfling JF, Anibal PC, Obando-Pereda GA, Peixoto IA, Furletti VF, Foglio MA, Goncalves RB. Antimicrobial potential of some plant extracts against Candida species. Braz J Biol. 2010;70:1065–8.

49. Ferrara AM. Treatment of hospital-acquired pneumonia caused by methicillin-resistant *Staphylococcus aureus*. Int J Antimicrob Agents. 2007;30:19–24.

50. Wenzel RP, Bearman G, Edmond MB. Community-acquired methicillin-resistant *staphylococcus aureus*(MRSA): New issues for infection control. Int J Antimicrob Agents. 2007;30:210–2.

51. Gould SW, Fielder MD, Kelly AF, Naughton DP. Anti-microbial activities of pomegranate rind extracts: Enhancement by cupric sulphate against clinical isolates of *S. aureus*, MRSA and PVL positive CA-MSSA.BMC Complement Altern Med. 2009;9:23.

52. Sharma M, Li L, Celver J, Killian C, Kovoor A, Seeram NP. Effects of fruit ellagitannin extracts, ellagic acid, and their colonic metabolite, urolithin A, on Wnt signaling. J Agric Food Chem. 2010;58:3965–9.

53. Afaq F, Zaid MA, Khan N, Dreher M, Mukhtar H. Protective effect of pomegranate-derived products on UVB-mediated damage in human reconstituted skin. Exp Dermatol. 2009;18:553–61.

54. Bae JY, Choi JS, Kang SW, Lee YJ, Park J, Kang YH. Dietary compound ellagic acid alleviates skin wrinkle and inflammation induced by UV-B irradiation. Exp Dermatol. 2010;19:e182–90.

55. Pai MB, Prashant GM, Murlikrishna KS, Shivakumar KM, Chandu GN. Antifungal efficacy of *Punica granatum*, *Acacia nilotica*, *Cuminum cyminum* and *Foeniculum vulgare* on *Candida albicans*: An *in vitro*study. Indian J Dent Res. 2010;21:334–6.

56. Bhadbhade SJ, Acharya AB, Rodrigues SV, Thakur SL. The antiplaque efficacy of pomegranate mouthrinse. Quintessence Int. 2011;42:29–36.

57. Menezes SM, Cordeiro LN, Viana GS. *Punica granatum* (pomegranate) extract is active against dental plaque. J Herb Pharmacother. 2006;6:79–92.

58. Promprom W, Kupittayanant P, Indrapichate K, Wray S, Kupittayanant S. The effects of pomegranate seed extract and beta-sitosterol on rat uterine contractions. Reprod Sci. 2010;17:288–96.

59. Kishore RK, Sudhakar D, Parthasarathy PR. Embryo protective effect of pomegranate (*Punica granatum*L.) fruit extract in adriamycin-induced oxidative stress. Indian J Biochem Biophys. 2009;46:106–11

60. Türk G, Sönmez M, Aydin M, Yüce A, Gür S, Yüksel M, et al. Effects of pomegranate juice consumption on sperm quality, spermatogenic cell density, antioxidant activity and testosterone level in male rats. Clin Nutr. 2008;27:289–96.

61. Hartman RE, Shah A, Fagan AM, Schwetye KE, Parsadanian M, Schulman RN, et al. Pomegranate juice decreases amyloid load and improves behavior in a mouse model of Alzheimer's disease. Neurobiol Dis.2006;24:506–15.

62. Dell'Agli M, Galli GV, Corbett Y, Taramelli D, Lucantoni L, Habluetzel A, et al. Antiplasmodial activity of *Punica granatum* L. fruit rind. J Ethnopharmacol. 2009;125:279–85.

63. Dell'agli M, Galli GV, Bulgari M, Basilico N, Romeo S, Bhattacharya D, et al. Ellagitannins of the fruit rind of pomegranate (*Punica granatum*) antagonize *in vitro* the host inflammatory response mechanisms involved in the onset of malaria. Malar J. 2010;9:208.

64. Neurath AR, Strick N, Li YY, Debnath AK. *Punica granatum* (pomegranate) juice provides an HIV-1 entry inhibitor and candidate topical microbicide. Ann N Y Acad Sci. 2005;1056:311–27.

65. Pirbalouti AG, Azizi S, Koohpayeh A, Hamedi B. Wound healing activity of Malva sylvestris and*Punica granatum* in alloxan-induced diabetic rats. Acta Pol Pharm. 2010;67:511–6.

66. Pirbalouti AG, Koohpayeh A, Karimi I. The wound healing activity of flower extracts of *Punica granatum* and Achillea kellalensis in Wistar rats. Acta Pol Pharm. 2010;67:107–10.

67. Hayouni E, Miled K, Boubaker S, Bellasfar Z, Abedrabba M, Iwaski H. Hydroalcoholic extract based-ointment from *Punica granatum* L. peels with enhanced *in vivo* healing

potential on dermal wounds.Phytomedicine. 2011;18:976–84.

68. Gil MI, Tomás-Barberán FA, Hess-Pierce B, Holcroft DM, Kader AA. Antioxidant activity of pomegranate juice and its relationship with phenolic composition and processing. J Agric Food Chem.2000;48:4581–9.

69. Tao X, Schulze-Koops H, Ma L, Cai J, Mao Y, Lipsky PE. Effects of Tripterygium wilfordii hook F extracts on induction of cyclooxygenase 2 activity and prostaglandin E2 production. Arthritis Rheum.1998;41:130–8.

70. Cerdá B, Cerón JJ, Tomás-Barberán FA, Espín JC. Repeated oral administration of high doses of the pomegranate ellagitannin punicalagin to rats for 37 days is not toxic. J Agric Food. 2003;51:3493–501.

71. Prescott SM, Fitzpatrick FA. Cyclooxygenase-2 and carcinogenesis. Biochim Biophys Acta.2000;1470:69–78.

72. Mena P, Girones-Vilaplana A, Moreno Diego A, García-Viguera C. Pomegranate fruit for health promotion: Myths and realities. Funct Plant Sci Biotechnol. 2011;5:33–42.

73. Shravan Kumar Y, Adukondalu D, Bhargavi Latha A, Vamshi Vishnu Y, Ramesh G, Shiva Kumar R, et al. Effect of pomegranate pretreatment on the oral bioavailability of buspirone in male albino rabbits. Daru.2011;19:266–9.

74. Nagata M, Hidaka M, Sekiya H, Kawano Y, Yamasaki K, Okumura M, et al. Effects of pomegranate juice on human cytochrome P450 2C9 and tolbutamide pharmacokinetics in rats. Drug Metab Dispos.2007;35:302–5.

75. Misaka S, Nakamura R, Uchida S, Takeuchi K, Takahashi N, Inui N, et al. Effect of 2 weeks' consumption of pomegranate juice on the pharmacokinetics of a single dose of midazolam:

An open-label, randomized, single-center, 2-period crossover study in healthy Japanese volunteers. Clin Ther. 2011;33:246–52.

76. Vidal A, Fallarero A, Peña BR, Medina ME, Gra B, Rivera F, et al. Studies on the toxicity of *Punica granatum* L. (*Punicaceae*) whole fruit extracts. J Ethnopharmacol. 2003;89:295–300.

4

CHAPTER

Heart Rot of Pomegranate: Disease Etiology and the Events Leading to Development of Symptoms

David Ezra, Benny Kirshner, Michal Hershcovich,

*Department of Plant Pathology and Weed Research, Agricultural Research Organization, the Volcani Center; and **Itzhak Kosto**, Extension Service, Ministry of Agriculture and Rural Development, Bet Dagan, 50250 Israel*

ABSTRACT

Pomegranate fruit rot disease, known as "heart rot" or "black heart," is a major pomegranate disease that impacts production worldwide. Heart rot is characterized by black rot of the fruit core that spreads from the calyx area, whereas the outer peel and the hard rind retain their healthy appearance. Aims of the present study were to identify the pathogen that causes pomegranate heart rot in Israel and to study the dynamics of fruit-organ colonization by fungi during fruit development, as the first steps toward the development of management strategies. Several fungi were isolated from pomegranate fruits, among which were species of Penicillium, Aspergillus,Botrytis, and Rhizopus, but the causative pathogen of the disease was identified asAlternaria alternata. This fungus was isolated from the pistils of more than 85% of open flowers and ~20% of the loculi of asymptomatic fruits and from all tissues of the symptomatic fruits. The dynamics of fruit-organ colonization by fungi during fruit development was recorded in six commercial orchards in 2010 and 2011. It was found that A. alternata spores germinate on the stigmata of open flowers and develop into the style. In some fruits, the mycelium grows into the tunnel and reaches the lower loculus, where the fungus becomes latent for about 3 to 4 months pending initiation of fruit ripening. Then, in a small number of fruits, the fungus starts growing and invades the arils, causing black rot of the arils in the lower loculus. At the beginning of disease development, the fungus causes brown soft rot of the arils, which becomes black and dry as the fungus grows. Eventually, the fungus grows from the lower loculus into the upper loculi, causing rot of the entire fruit.

Pomegranate (*Punica granatum* L., Punicaceae) is an ancient plant and fruit native to central Asia. The plant is a multitrunk shrub with bushy appearance, which under domestication is grown as a tree up to 5 m in height (Levin 2006), of which most varieties are deciduous. Because the pomegranate tree is highly adaptive to a wide range of climatic and soil conditions, it is grown in many different geographical regions including the Mediterranean basin, Asia, and California (Holland et al. 2009). Pomegranate fruit may be consumed as fresh fruit, juice, or wine; currently, separated arils are sold as a product in supermarkets.

Growing awareness of the medical and health benefits of the pomegranate fruit has stimulated a sharp increase in consumption of the fruit and juice and, as a result, there has been significant growth in orchard planting worldwide. New orchards have been planted in the traditional growing regions, in California, and in the southern hemisphere in South America, South Africa, and Australia (Day and Wilkins 2009; Holland et al. 2009). Recent scientific findings indicate that tissues of the pomegranate fruit, flowers, bark, and leaves contain bioactive phytochemicals that are antimicrobial, reduce blood pressure, and are active against serious diseases such as diabetes and cancer (Seeram et al. 2006). Among pomegranate's longest known health benefits are its activities against infectious diseases (Jayaprakasha et al. 2006), and antimicrobial activity of pomegranate extracts has been demonstrated against a wide array of bacteria, fungi, and viruses (Jayaprakasha et al. 2006).

In Israel, until the year 2000, pomegranate was considered a minor crop grown on about 150 ha. However, as a result of research emphasizing the health benefits of the fruit, the extent of tree planting rose dramatically: with the addition of new varieties, it stood at 400 ha by 2004, and by 2013 the extent of plantations was estimated at 2,700 ha. Annual production is about 60,000 tons of fresh fruit, of which about 50% is exported. The main varieties grown in Israel are: 'Wonderful' (red, sweet-sour, occupying about 80% of the production area); 'Emek,' 'Akko,' and 'Shani' (pink and red, sweet, early ripening varieties, occupying about 12% of the production area); and '116,' 'Hershkowitz,' and 'Camel' (red, sour, occupying about 8% of the production area) (Israeli fruit board and I. Kosto, *personal communication*). Flowering usually begins at the beginning of May-June. Fruit set and growth lasts until the end of the summer (August for the early varieties 'Emek,' 'Akko,' and 'Shani;' September for '116,' 'Hershkowitz,' and 'Camel;' and October for 'Wonderful') (Holland et al. 2009 and I. Kosto, *personal communication*).

Pomegranate fruit rot disease, known as "heart rot" or "black heart," is a major pomegranate disease that impacts production worldwide (Day and Wilkins 2009; Holland et al. 2009; Pala et al. 2009; Tziros et al. 2008). When infected fruits are opened, brown (soft) to black (dry) rot of the arils is observed. Nevertheless, the outer peel and the hard rind of infected fruits retain their healthy appearance (Tziros et al. 2008). Fruits with heart rot can be identified in the orchard according to their skin color and their light weight; furthermore, when such fruits are knocked, they emit a hollow sound, in contrast to the dull sound from healthy fruits. There is, at present, no effective control in the orchard and in many cases, healthy-looking fruits develop the disease in storage. Fruit loss may be substantial, as was reported, for example, from California (Zhang and McCarthy 2012) and Greece, and may cause significant damage of up to 50% of the fruits in the orchards (Tziros et al. 2008). However, one long-term damaging effect of the disease concerns the quality of the product and its reputation and a single diseased fruit purchased by a consumer might cause individual and his relatives to reject consuming pomegranate fruits for years to come. Another problem can arise in industry where a diseased fruit entering a juice or fresh aril production process might cause spoilage or delay production.

In most cases, *Alternaria* spp. and *Aspergillus* spp. were isolated from diseased pomegranate fruits (Michailides et al. 2008). Other fungi (such as *Penicillium* sp., *Botrytis*sp., and *Rhizopus* sp.) were also isolated. Accordingly, it is not always clear which is/are the causal agent(s) of the disease and what is the role of the other fungi in the disease syndrome. It is generally accepted that the infection of fruits occurs in the orchard during flowering (Zhang and McCarthy 2012; Ezra, *unpublished data*), but experimental data addressing the infection process and disease development in the fruit are currently unavailable. *Alternaria alternata* (Fr.) Keissl. may also infect other tissues of the pomegranate tree (Ezra et al. 2010; Pala et al. 2009; Tziros et al. 2008). In Turkey, *A. alternata* caused lesions on leaves, flowers, and young fruits (Pala et al. 2009). Recently, we reported on the appearance of novel symptoms of *A. alternata* infection in pomegranate orchards in Israel (Ezra et al. 2010; Gat et al. 2012): black spots on leaves and fruit surface, as well as chlorosis, and abscission of some leaves. On fruits, the damage was restricted to the peel surface, with the edible tissue remaining unaffected. This disease was later reported from Spain and Egypt as well (Ammar and El-Naggar 2014; Berbegal et al. 2014).

Two main scenarios represent possible descriptions of heart rot development in the orchard. According to the first scenario, a spore penetrates the pistil of an open flower and grows into the tunnel (an opening connecting the pistil to the loculus) and then into the loculus, where it remains latent until the ripening fruit can support its growth. Later, it resumes its growth in the lower loculus, causing rot of the arils as it grows into the upper loculus and eventually infects the entire fruit. According to the second scenario, the infection of the pistil occurs at some point in the course of fruit growth. The spore then germinates, penetrates the pistil, grows through the tunnel into the lower loculus, and causes rot of the arils as it grows into the upper loculus, eventually infecting the entire fruit. If the first scenario is correct, it should be possible to isolate the rot-causing fungus from the pistil, tunnel, and loculus at all fruit-development stages, from flower opening to fruit ripening. However, according to the second scenario, it should not be possible to isolate the fungus from any of the fruit parts during the first few stages of fruit development, but during later stages of development the isolation of the fungus increases abruptly and then remains steady until fruit ripening. Our research hypothesis was that identifying the causal agent of the heart rot disease and determining the chain of events leading to development of symptoms within the fruits would enable us to formulate concepts applicable to the disruption of these processes. This would lead, in turn, to development of means for disease suppression, prevention of yield losses, and improvement of the quality of pomegranate products. The specific objectives of the present study were: (i) to identify the pathogen that causes pomegranate heart rot in Israel, and (ii) to study the dynamics of fruit-organ colonization by fungi during fruit development.

MATERIALS AND METHODS

Fungal isolation, identification, and pathogenicity tests.

Fruits suspected to be diseased and fruits that were presumed healthy (35 fruits from each group) were collected in 2009 from commercial pomegranate orchards, starting one month before ripening (Table 1). Suspected fruits exhibited the following symptoms: discoloration (redness) of one part of the fruit; emission of a hollow sound when knocked, in contrast to the dull sound from healthy fruits; easy release from the tree when picked; light weight; and in some cases, a rugged peel. The fruits were taken to the laboratory, surface-sterilized in 3% sodium hypochlorite solution for 1 min, and rinsed twice in distilled water. The fruits were cut with a sterile knife and then divided into two groups: those that exhibited any internal symptoms (such as discoloration or rot) and those that were asymptomatic. Rotten aril tissues from inside the fruit were placed on potato dextrose agar (PDA) (Acumedia, Lansing, MI) amended with tetracycline (Sigma, Rehovot, Israel) at 12 μg ml^{-1}; the plates were incubated at 25°C. Mycelium from fungal colonies was transferred to new plates and left to grow for 5 to 7 days prior to macroscopic and microscopic identification. Single-spore fungal isolates were prepared and used in the artificial inoculation experiments. As comparable symptoms were observed in all orchards and for the two varieties tested, data presented in this report were pooled. Healthy looking arils from nonsymptomatic fruits were used for *Penicillium* isolation as described above.

Table 1. Details of the orchards from which fruits were sampled in this study

Village	Latitude	Longitude	Cultivar	Sampling season
Kidron	N31°49′13.44″	E34°48′4.32″	Wonderful	2009, 2010, 2011
	N31°49′13.44″	E34°48′4.32″	116	2009, 2010
	N31°49′39.36″	E34°46′20.64″	Akko	2011
Givat Nili	N32°32′51.36″	E35°02′45.6″	Wonderful	2009, 2010, 2011
Amikam	N32°34′35.04″	E35°02′19.68″	116	2009, 2010, 2011
Ein Iron	N32°29′44″	E35°01′27.84″	Akko	2011

Koch's postulates with both *Alternaria* sp. and *Penicillium* sp. (two independent experiments, using seven different isolates for *A. alternata* and three isolates of*Penicillium* sp.) were tested with detached fruits. The fruits were surface sterilized as described above, cut horizontally, and inoculated with a 10 µl drop of 10^6 conidia ml^{-1}suspension from each isolate. Alternatively, instead of using a spore suspension, a fresh plug of 1-week-old fungal culture grown on PDA was placed in the center of a cut fruit. The inoculated fruits were placed in sealed 1-liter plastic boxes and incubated for 1 week at 25°C. After symptoms appeared, the fungus was reisolated from infected fruits as described above. Fungi isolated according to Koch's postulates were transferred to new plates and left to grow for 5 to 7 days pending macroscopic and microscopic identification. Additionally, single-spore colonies were used for fungal DNA isolation according to Gat et al. (2012), and were used for amplifying the internal transcribed spacer (ITS)-5.8S rDNA, elongation factor (EF-1, alpha) and glyceraldehyde-3-phosphate dehydrogenase (gpd) sequences. The purified products were sent for direct PCR sequencing (Macrogen, Geumcheon-gu Seoul, South Korea). Sequences were submitted to GenBank and were deposited as accession numbers KC354713, KM052593 to KM052598, KJ830919 to KJ830925, and KJ830926 to KJ830931.

Fruit sampling for disease etiology.

During 2010 we collected fruits of two pomegranate varieties—the late-season 'Wonderful' and the early-season '116'—from the central (Kidron) and the northern (Givat Nili and Amikam) regions of Israel. In 2011, 'Wonderful,' '116,' and an additional variety—the early-season 'Akko'—were collected from Kidron, Givat Nili, and Amikam, and from Ein Iron (Table 1). In each orchard, 20 trees were randomly selected and marked with colored bands. In 2010, flowers and fruits were sampled from each tree every 2 weeks, whereas in 2011, plant organs were sampled at several different growth stages: closed flower, open flower, fruit-bearing, small fruits, and ripe fruits.

At each sampling date, the crop developmental stage was recorded according to the BBCH General Scale (Melgarejo et al. 1997), as follows: 51 = appearance of flower buds; 55 = swollen calyx; 61 = open flower; 71 = young fruit; 73 = developing fruit; and 81 = fruit ripening. One fruit or flower was randomly picked from the perimeter of the tree at each collection time. The samples were surface-sterilized in 3% sodium hypochlorite solution for 1 min and rinsed twice in distilled water. The flowers or fruits were cut open with a sterile knife, and the pistil was cut into five pieces, which were placed onto PDA amended with tetracycline (Sigma; Rehovot, Israel) at 12 μg ml^{-1} in the same relative positions as in the intact pistil. The same treatment was applied to the fruit tunnel. Arils (at least four) were sampled from the loculi and placed on PDA. The plates were incubated at 25°C. Mycelium from fungal colonies was transferred to new plates and left to grow for 5 to 7 days pending macroscopic and microscopic identification. All sampled fruits and plates were documented. The percent incidence of colonized pistils, fruit tunnels, and arils for each fruit and the percent incidence of symptomatic fruits were calculated for each sampling date, for each orchard. As similar trends were observed in samples taken from both of the cultivars tested in 2010 and from all three cultivars tested in 2011, the data were pooled. The incidence of colonized or diseased organs (± standard error) was calculated for each sampling date or crop developmental stage.

Data analysis.

Statistical analyses were performed using JMP software (SAS Institute Inc.). Means and standard error values were computed for data recorded in 2010 and 2011 for studying the dynamics of fruit-organ colonization by fungi. Differences in the incidence of stained and unstained tunnels in symptomatic and asymptomatic pomegranate fruits were determined according to the χ^2 test at $P \leq 0.05$.

RESULTS

Identification of the pathogen causing pomegranate heart rot.

In the orchards, fruits with moldy arils could be categorized into two groups according to the symptoms observed on the fruit surface. The first group included fruits exhibiting noticeable external damage on the peel surface—damage that could have resulted from insects invading the mesocarp, rupture of the peel by physical impacts, e.g., a thorn or adjacent branch of the tree, or some other cause. In such cases, fungi had invaded the fruits through the wounds, and rotted arils were scattered in the vicinity of the wounds. Occasionally, symptoms were also visible on the peel surface. For example, the damage presented in Figure 1A is probably due to a mealybug (*Pseudococcus maritimus*). Damage in the crown enabled penetration *of Aspergillus* spp. into the fruit, causing rotting of the mesocarp tissue (Fig. 1B); in the case of penetration by *Aspergillus* spp., no aril rot was observed. These symptoms were not regarded as heart rot. The second group included fruits that did not exhibit any noticeable damage on the peel surface (Fig. 1C); when these fruits were cut, some or all of the arils were rotten and fungal mycelium was observed within the fruits. It should be noted that some of the arils exhibited brown (soft) or black (dry) rot, presumably due to differences in rot development (Fig. 1D). No rotting of the mesocarp was observed in these fruits. In our present study these symptoms were regarded as heart rot.

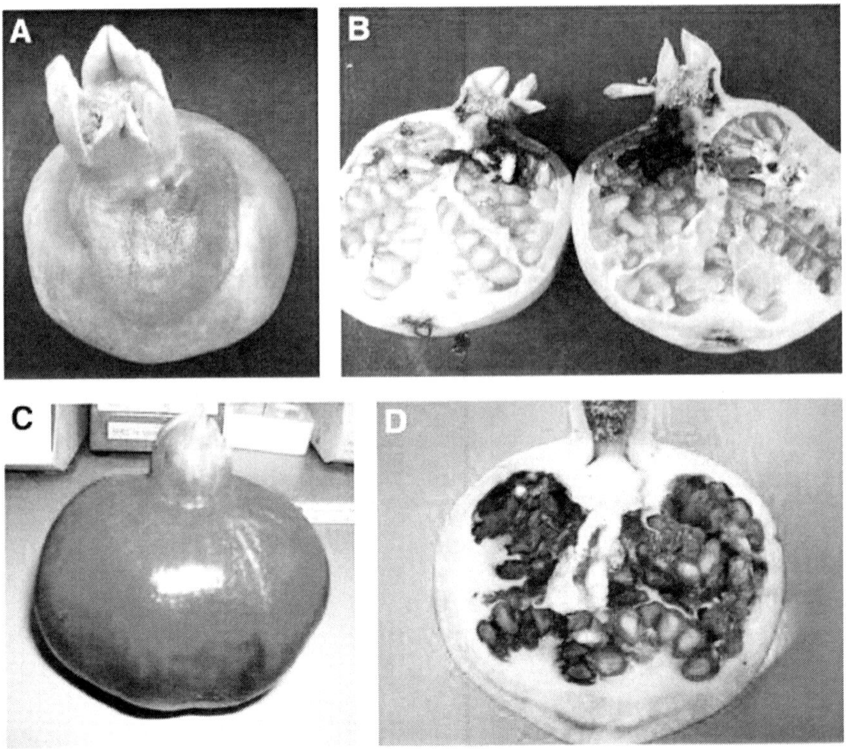

Fig. 1. External and internal symptoms on pomegranate fruits. **A** and **B,** Rot caused by *Aspergillus* sp. It was not considered as heart rot in the present study. **C**and **D,** Rot caused by *Alternaria alternata*; external symptoms were not observed but the internal part of the fruit, including the arils and the inner lamella, exhibited brown (soft) or black (dry) rot. These types of symptoms were considered as heart rot in the present study.

Developed (suspected and nonsuspected) fruits were collected in the orchards and cut longitudinally through the pistil, which made it possible to view the tunnel leading from the pistil to the lower loculus of the fruit (Fig. 2A). The tunnel remained open until the fruit ripened. Occasionally, fungal mycelia and sporulation were clearly observed on the stamens that remained attached during the advanced stages of fruit development (Fig. 2B). The moldy stamens were located in the vicinity of the stigma and, in some fruits, the inner part of the pistil and the upper portion or the whole of the tunnel were stained. These stained tunnels were either brown or black (Fig. 2C and D), and mycelia were occasionally observed in them (Fig. 2E). Staining of the upper portion of the tunnel was observed in 29.9% of the asymptomatic fruits, i.e., those in which no rotten arils were observed when the fruit was cut open. In 75.2% of the symptomatic fruits, rot was restricted to a small number of arils; in 95.4% of these, the rotten arils were located in the lower loculus, close to the calyx (Fig. 2F). The tunnels of all symptomatic fruits were stained (Fig. 3A). Attempts to isolate fungi from white and discolored tunnels of mature fruits revealed that *Penicillium* sp. colonized 70 and 75% of these fruits, respectively, and*Alternaria* sp. were isolated from 26.9 and 100% of them, respectively (Fig. 3B).

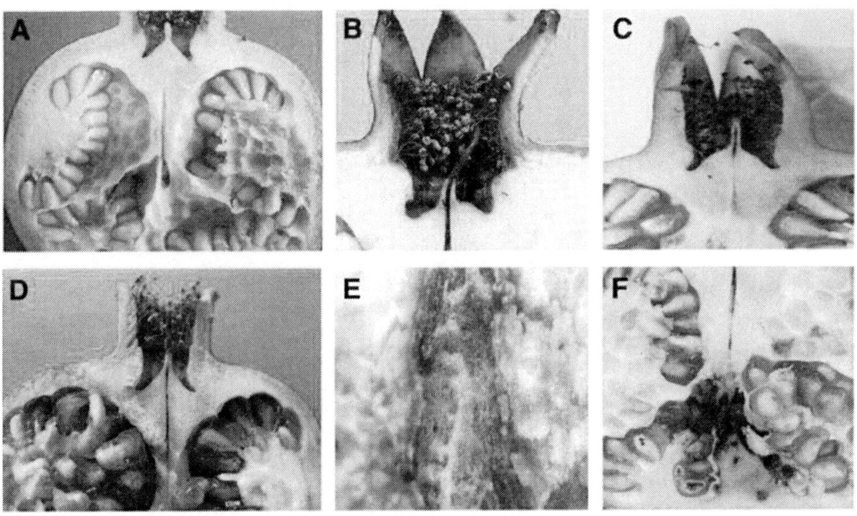

Fig. 2. Longitudinal cuts of asymptomatic and symptomatic pomegranate fruits. **A,** The tunnel leading from the pistil to the lower loculus of the fruit. **B,** Mycelia and sporulation of fungi developing on the stamens. **C** and**D,** Discoloration of the inner part of the pistil and the tunnel. **E,** Mycelia developing in stained tunnels. **F,**Rot initiation in the lower loculus of the fruit, close to the calyx.

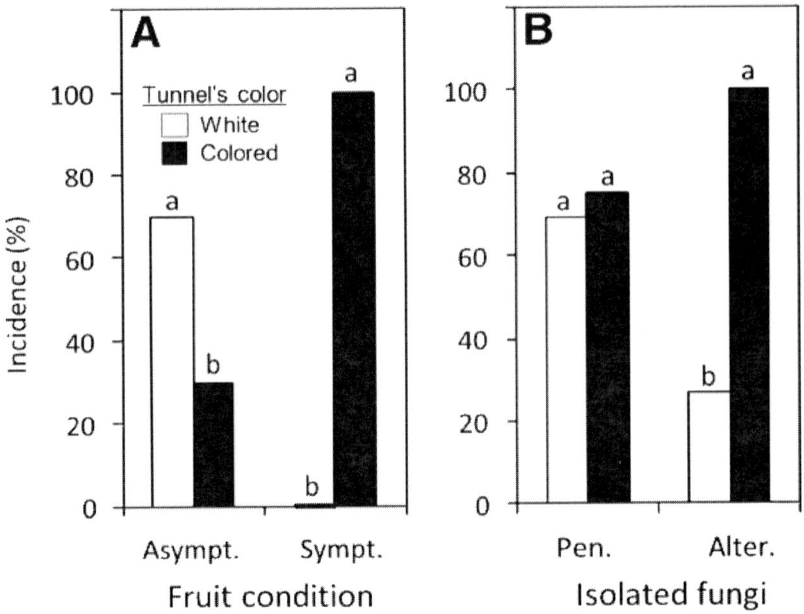

Fig. 3. **A,** Incidence of unstained and stained tunnels in asymptomatic and symptomatic pomegranate fruits.**B,** Incidence of fruits with white and stained tunnels, from which *Penicillium* spp. or *Alternaria* spp. were isolated. For each group in each graph, values of bars followed by different letters differed significantly at $P \leq 0.05$, according to the χ^2 test.

In the light of macroscopic and microscopic examination of the characteristics of mycelia and spores developing in vitro, the fungus isolated from both soft and dry rotten arils was identified as a species of *Alternaria*. Several single spore *Alternaria* cultures were further examined through sequencing of ITS-5.8S rDNA, elongation factor (EF-1, alpha), and glyceraldehyde-3-phosphate dehydrogenase (gpd). All seven isolates used in this study exhibited at least 99% identity to sequences of *A. alternata* (Nees:Fr.) Keissl available in GenBank. Artificial inoculation of fruits resulted in black rot of the arils, and *A. alternata*could be reisolated from the rotten arils. All inoculations of detached fruits with *Penicillium*spp. isolated from the inner parts of fruits collected in the orchards did not result in black typical rot of the arils. In some cases, local collapse of the arils with slight change in aril color (light brown) could be observed in the inoculation site (data not shown). The ITS-5.8S rDNA sequence of a *Penicillium* isolate used in this study was found to have 100% identity to *Penicillium adametzioides*. Since the other *Penicillium* spp. isolated failed to cause fruit rot under laboratory conditions, they were not fully identified to the species level.

The dynamics of fruit-organ colonization by fungi during fruit development.

The dynamics of fruit-organ colonization by fungi during fruit development was recorded in 2010 and 2011. In 2010, fungi were rarely isolated from closed flowers (results not shown), but >75% of the pistils collected from the early stages of fruit development until maturity were colonized with fungi (Fig. 4A). These fungi were identified as *Penicillium*spp., *Aspergillus* spp., *A. alternata*, *Botrytis* spp., and *Rhizopus* spp. Other fungal species were isolated in only a few samples, mainly in the early stages of fruit development, so they were not identified. The incidence of colonized tunnels and loculi was low in the early stages of fruit development, but increased over time. The tunnels and loculi of more than 85% of the fruits were colonized by fungi (Fig. 4A), with the colonization rate in the former being 10 to 40% higher than the latter. Two fungi predominated as time progressed: *A. alternata* colonized the pistils, and *Penicillium* spp. colonized the tunnels and the loculi (Fig. 4B and C). The proportion of fruits from which *Penicillium* spp. was isolated from loculi increased gradually with time (and fruit maturity), and was recovered from 87% of the loculi of mature fruits. On the other hand, colonization of the loculi by *A. alternata* was sporadic and did not change much as fruit matured. Out of a total of 666 sampled fruits, only 18 (2.7%) exhibited the typical heart rot symptoms, and out of the 222 fruits sampled from late-July onwards, representing growth stages 73 to 81, the incidence of symptomatic fruits was 8.1%.

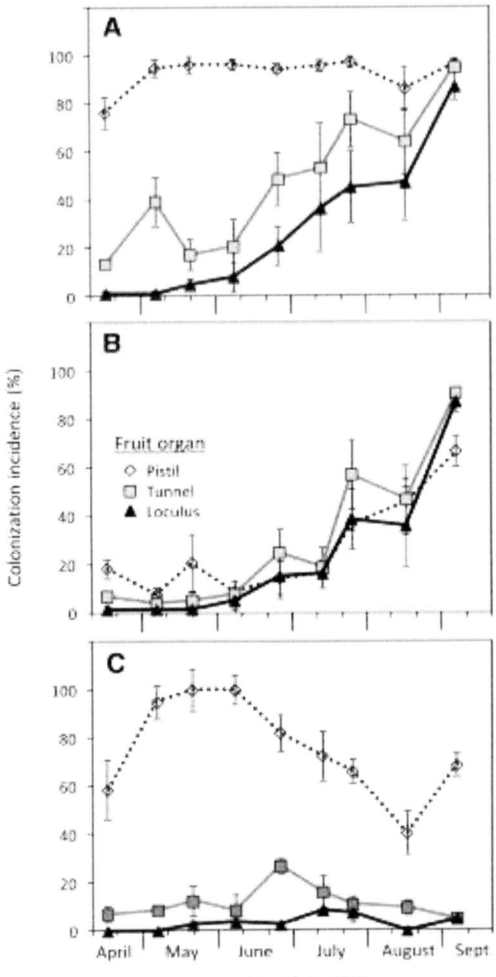

Fig. 4. Dynamics of fruit-organ colonization by fungi during fruit development in 2010. **A,** Fungal colonization (by any species) of the pistil, the tunnel and the lower loculus. **B,** Colonization by*Penicillium* spp. of fruit organs. **C,** Colonization by *Alternaria* spp. of fruit organs. Vertical bars represent the standard error of the mean.

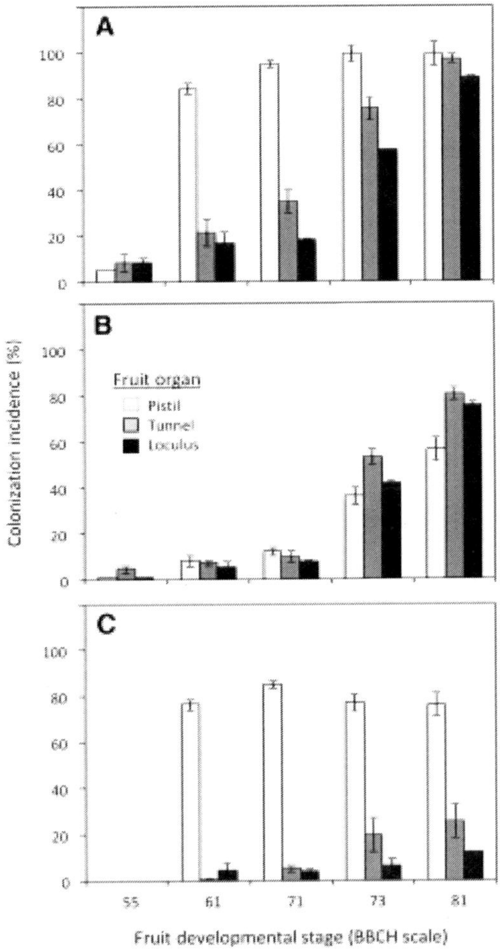

Fig. 5. Dynamics of fruit-organ colonization by fungi during fruit development in 2011. **A,** Fungal colonization by any species in the pistil, the tunnel, and the lower loculi. **B,** Colonization by*Penicillium* spp. of fruit organs. **C,** Colonization by *Alternaria* spp. of fruit organs. Fruit developmental stages according to BBCH General Scale: 51 = appearance of flower buds; 55 = swollen calyx; 61 = open flower; 71 = young fruit; 73 = fruit development; and 81 = fruit ripening. Vertical bars represent the standard error of the mean.

In general, results of the sampling in 2011 resembled those obtained in 2010; the predominant fungi were *Penicillium* sp. and *A. alternata*, with the former being more abundant in the tunnels and loculi and the latter in the pistils (Fig. 5). Although 75.8% of loculi of mature fruits were colonized by *Penicillium* sp., only 12 (2%) of the sampled fruits exhibited the typical heart rot symptoms. Among the 240 fruits sampled during growth stages 73 to 81, the incidence of symptomatic fruits was 5%.

DISCUSSION

The specific objectives of the present study were to identify the pathogen(s) causing pomegranate heart rot in Israel and to study the dynamics of fruit-organ colonization by fungi during fruit development. It was previously reported that the causal agent of heart rot disease of pomegranates was *A. alternata* (Day and Wilkins 2009; Holland et al. 2009;Tziros et al. 2008; Zhang and McCarthy 2012). In Israel, it was commonly agreed that pomegranate heart rot was caused by a complex of fungi—among which were *Alternaria*spp. and *Aspergillus* spp. (Ezra; *unpublished data*). The present results suggest that in Israel, *A. alternata* (Nees:Fr.) Keissl is the predominant pathogen causing pomegranate heart rot.

Two main scenarios might describe the dynamics of pomegranate fruit-organ colonization by the pathogen during fruit development. According to the first, spores infect the pistils of open flowers; they grow into the tunnel and then into the loculus, where they remain latent until the fruits reach a ripening stage that supports their development. The pathogen then colonizes the lower loculus, causing rot of the arils, and later invades the upper loculi and, eventually, the entire fruit. According to the second scenario, infection of the pistil occurs some time during fruit growth, when the fruit internal environment can support infection and fungal growth.

The pathogen develops into the pistil, grows through the tunnel into the lower loculus, where it rots the arils, and then proceeds into the upper loculus, eventually infecting the entire fruit.

Our present results demonstrate that the fruits were infested with *A. alternata* from flower opening to fruit ripening; this pathogen was isolated from the pistils of more than 80% of the sampled fruits at all stages from flower opening to fruit ripeness. It was isolated at very low incidence from the tunnels and loculi of asymptomatic fruits, but from all the black tunnels of symptomatic fruits (Fig. 3B). This implies, presumably, that the first of the above scenarios better represents the etiology of pomegranate heart rot in Israel. On the other hand, infection of the fruits by *Penicillium* followed the second of the suggested scenarios: isolation of this fungus from flowers and small fruits—growth stages 71 to 73—was very low, but there was a gradual rise in fungal isolation from fruits at growth stages 73 to 81, and the increase continued as the fruits ripened. It is worth mentioning that no obvious symptoms attributable to the presence of this fungus were observed in the orchard or the laboratory performing Koch's postulates. Yet, it was previously reported that*Penicillium* spp. (*P. glabrum*, *P. implicatum*, and others) cause rot of the arils and the peel of stored pomegranate fruit (Bardas et al. 2009; Labuda et al. 2004; Palou et al. 2010,2013). The internal rot was described as soft, brown tissue that became covered with green mycelium, while the calyx showed green conidiophores (Bardas et al. 2009). The Penicillium rot described in these studies resembles the collapse of arils and color change we observed in our attempt to inoculate detached fruit with *Penicillium* spp. isolated from internal parts of fruit, still different from the black rot typical to heart rot caused by *A. alternata* in pomegranate both in texture and color.

The above discussion implies that suppression of heart rot incidence by preventing *A. alternata* from infecting the flowers (e.g., by application of fungicides) is impractical. The fungus penetrates its host through the stigmata and this is the plant organ to be protected by the fungicides. Directing fungicides to the stigmata of pomegranate flowers is a difficult task in view of the position of this organ in the developing flower and the erratic orientation of the flowers in the tree canopy. Furthermore, during the blooming period, new flowers open daily and flowers persist in the orchard for several weeks. This implies that fungicides, if deposited in the infection site (i.e., the stigmata), should be applied in short intervals for a long period of time. Indeed, attempts to suppress pomegranate heart rot in Israel by fungicide application resulted in insufficient results (I. Kosto, *unpublished data*) and this measure is no longer used by pomegranate growers.

Still unsolved is the question of why very few fruits developed rot whereas most of them were colonized with *A. alternata* and probably had the potential to develop the disease. It is possible that fruits located at different positions on the tree were exposed to slightly differing environmental conditions—e.g., sunlight, shade, etc.—and that these differences affect pathogen invasion and development. Differences in disease incidence may be explained by differences among cultivars as described by Palou et al. (2013) who found differences in Alternaria infection of cv. Wonderful vs. cv. Mollar de Elche growing in the same area in Spain (Palou et al. 2013). Yet, in our study we used commercial cultivars grown in Israel and among those we did not find fundamental differences in percent incidence of symptomatic fruits. Alternatively, it is possible those factors related to the host affect fruit physiology and that individual fruits differ in their susceptibility to pathogen invasion and rot development. This might happen, for example, if translocation of essential factors and elements were associated with fruit capability to resist disease development. Recently, it was demonstrated that the development of core rot caused by *A. alternata* in 'Red Delicious' apples was governed by host physiology rather than pathogen occurrence or environmental conditions at the bloom stage (Shtienberg 2012); it was found that the incidence of core rot was higher in light-yielding trees than in heavy-yielding ones and vice versa. The reasons for these findings are not yet known; however, light cropping in apple coincided with excessive shoot growth, increased fruit size, and lower fruit calcium concentration. The reduced calcium concentration in fruits growing on trees with excessive shoot growth has been attributed to stronger competition for available calcium between fruits, on the one hand, and vigorous new shoots with heavy foliage, on the other hand, at the most critical time for fruit development (Greene and Lord 1983). In heavy-cropping trees, fruits were usually relatively small but had high calcium concentrations which, on average, declined in inverse proportion to increasing fruit size; calcium concentration in apples from light-yielding trees was lower in all fruit-size classes than that in apples from trees with heavy crop loads (Sharples 1974; Shtienberg 2012). Calcium is considered to be an inhibitor of polygalacturonase, an enzyme produced by *A. alternata*during tissue invasion by the fungus. This enzyme, among other pectolytic enzymes produced by parasitic fungi including *A. alternata*, is needed by the fungus to promote rot development (Bateman and Lumsden 1965; Elad and Evensen 1995; Taiz and Zeiger 2002). These findings suggest that these correlations may be part of a possible explanation of the disease occurrence in apples; they also present a different aspect of the disease development and a possible role of plant physiology in pomegranate heart rot disease.

ACKNOWLEDGMENTS

This work was financed by the Chief Scientist of The Ministry of Agriculture and Rural Development (MOARD) of Israel; grant number 132-1534-11. The authors would like to thank Mrs. Reoma Lambrez of the Ministry of Agriculture and Rural Development Extension Service for her assistance, and Ronen Ilel, Itzik Alfasi, Edi Benedicts, Mahamod Jorban, Gori Rozen, and Yaniv Aboksis for letting us use their orchards for these experiments. Contribution No. 536/13 from the Agricultural Research Organization, The Volcani Center, Bet Dagan, Israel.

Literature Cited

Ammar, M. I., and **El-Naggar, M. A.** 2014. Screening and characterization of fungi and their associated mycotoxins in some fruit crops. Inter. J. Adv. Res. 2:1216-1227.

Bardas, G. A., Tzelepis, G. D., Lotos, L., and **Karaoglanidis, G. S.** 2009. First report of *Penicillium glabrum* causing fruit rot of pomegranate (*Punica granatum*) in Greece. Plant Dis. 93:1347. 10.1094/PDIS-93-12-1347A

Bateman, D. F., and **Lumsden, R. D.** 1965. Relation of calcium content and nature of pectic substances in bean hypocotyls of different ages to susceptibility to an isolate of*Rhizoctonia solani*. Phytopathology 55:734-738.

Berbegal, M., López-Cortés, I., Salazar, D., Gramaje, D., Pérez-Sierra, A.,García-Jiménez, J., and **Armengol, J.** 2014. First report of Alternaria black spot of pomegranate caused by *Alternaria alternata* in Spain. Plant Dis. 98:689. 10.1094/PDIS-07-13-0717-PDN

Day, K. R., and **Wilkins, E. D.** 2009. Commercial pomegranate (*Punica granatum* L.) production in California. Acta Hortic. 890:275-286.

Elad, Y., and **Evensen, K.** 1995. Physiological aspects of resistance to *Botrytis cinerea*. Phytopathology 85:637-643.

Ezra, D., Gat, T., Skovorodnikova, Y., Vardi, Y., and **Kosto, I.** 2010. First report of Alternaria black spot of pomegranate caused by *Alternaria alternata* in Israel. Australas. Plant Dis. 5:1-

2. 10.1071/DN10001

Gat, T., Liarzi, O., Skovorodnikova, Y., and **Ezra, D.** 2012. Characterization of*Alternaria alternata* causing black spot disease of pomegranate in Israel using a molecular marker. Plant Dis. 96:1513-1518. 10.1094/PDIS-12-11-1041-RE

Greene, D. W., and **Lord, W. J.** 1983. Effects of dormant pruning, summer pruning, scoring, and growth-regulators on growth, yield, and fruit-quality of Delicious and Cortland apple-trees. J. Am. Soc. Hortic. Sci. 108:590-595.

Holland, D., Hatib, K., and **Bar-Ya'akov, I.** 2009. Pomegranate: botany, horticulture, breeding. Hortic. Rev. (Am. Soc. Hortic. Sci.) 35:127-191.

Jayaprakasha, G. K., Negi, P. S., and **Jena, B. S.** 2006. Antimicrobial activities of pomegranate. Pages 167-183 in: Pomegranates: Ancient Roots to Modern Medicine. **N. P. Seeram, R. N. Schulman**, and **D. Heber**, eds. CRC Press Taylor & Francis Group,Boca Raton, FL.

Labuda, R., Hudec, K., Piecková, E., Mezey, J., Bohovic, R., Mátéová, S., and**Lukác, S. S.** 2004. *Penicillium implicatum* causes a destructive rot of pomegranate fruits. Mycopathologia 157:217-223. 10.1023/B:MYCO.0000020599.95040.c6

Levin, G. M. 2006:Pages 15-183 in: Pomegranate roads: a Soviet botanist's exile from Eden. **B. L. Baer**, ed. Floreat Press, Forestville, CA.

Melgarejo, P., Martinez-Valero, R., Guillamon, J. M., Miro, M., and **Amoros, A.**1997. Phenological stages of the pomegranate tree (*Punica granatum* L.). Ann. Appl. Biol. 130:135-140. 10.1111/j.1744-7348.1997.tb05789.x

Michailides, T., Morgan, T., Quist, M., and **Reyes, H.** 2008. Infection of pomegranate by *Alternaria* spp. causing black heart. Phytopathology 98:S105.

Pala, H., Tatli, A., Yilmaz, C., and **Özgüven, A. I.** 2009. Important diseases of pomegranate fruit and control possibilities in Turkey. Acta Hortic. 818:285-290.

Palou, L., Guardado, A., and **Montesinos-Herrero, C.** 2010. First report of*Penicillium* spp. and *Pilidiella granati* causing postharvest fruit rot of pomegranate in Spain. New Dis. Rep. 22:21. 10.5197/j.2044-0588.2010.022.021

Palou, L., Taberner, V., Guardado, A., del Río, M. A., and **Montesinos-Herrero, C.**2013. Incidence and etiology of postharvest fungal diseases of pomegranate (*Punica granatum* cv. Mollar de Elche) in Spain. Phytopathol. Mediterr. 52:478-489.

Seeram, N. P., Zhang, Y., Reed, J. D., Krueger, C. G., and **Vaya, J.** 2006. Pomegranate phytochemicals. Pages 3-29 in: Pomegranates: Ancient Roots to Modern Medicine. N. P. Seeram, R. N. Schulman, and D. Heber, eds. CRC Press Taylor & Francis Group, Boca Raton, FL.

Sharples, R. O. 1974. The number and size of fruits in relation to rootstock type, application of growth regulators, pruning, and fruit thinning. Acta Hortic. 45:21-24.

Shtienberg, D. 2012. Effects of host physiology on the development of core rot, caused by *Alternaria alternata*, in Red Delicious apples. Phytopathology 102:769-778.10.1094/PHYTO-09-11-0260

Taiz, L., and **Zeiger, E.** 2002. Mineral nutrition. Pages 67-86. in: Plant Physiology, 3rd Ed., **Taiz, L.**, and **Zeiger, E.**, eds. Sinauer Associates, Inc., Sunderland, MA.

Tziros, G. T., **Lagopodi, A. L.**, and **Tzavella-Klonari, K.** 2008. *Alternaria alternata*fruit rot of pomegranate (*Punica granatum*) in Greece. Plant Pathol. 57:379.10.1111/j.1365-3059.2007.01668.x

Zhang, L., and **McCarthy, M. J.** 2012. Black heart characterization and detection in pomegranate using NMR relaxometry and MR imaging. Postharvest Biol. Technol. 67:96-101.

5

CHAPTER

Relationships between Fruit Acceptability and Health-Case of Seven Pomegranate (Punica granatum L.) Juices

Chiraz El Kar[1*], Narjes Mtimet[2], Ali Ferchichi[1], Jalloul Bouajila[2*]

[1]Laboratoire d'Aridoculture et de Cultures Oasiennes, Institut des Régions Arides, Médenine, Tunisie; [2]Université de Toulouse, Laboratoire des Interactions Moléculaires et Réactivité Chimique et Photochimique UMR CNRS 5623, Université Paul-Sabatier, Toulouse, France

ABSTRACT

The objective of this study was to evaluate acceptability and instrumental measurements (titratable acidity, total soluble sugar, total phenolic, flavonoids, anthocyanins, and condensed tannin contents as well as antioxidant capacity) of seven Tunisian pomegranate cultivars to relate fruit acceptability and beneficial health. Seventy one participants were asked to complete a questionnaire and to indicate their level of acceptance, on a nine point-hedonic scale, for fruit size, sweetness, bitterness, juice color and juice overall acceptability. Taste and aril color were the main criteria to select Pomegranate. Significant differences occurred for all the acceptability tests. Titratable acidity determines the taste and important inter-specific variability in the phenolic contents as well as for the antioxidant capacities was noticed. Anthocyanin and condensed tannin were contributing to the organoleptic properties while bitterness acceptability was negatively correlated to the antioxidant capacity. Also, a relationship was found between the fruit size acceptability and the antioxidant capacity.

Keywords: Pomegranate Juice; Acceptability Tests; Instrumental Measurements; Phenolic Contents; Antioxidant Capacity

1. INTRODUCTION

"The eyes are the gatekeeper to the mouth". Color was suggested to be the most influential quality factor, as consumers have expectations of overall quality based on color [1]. Most of the yellow, red and blue colors in fruit are related to the flavonoids [2]. Flavonoids are a group of polyphenolic compounds naturally present in a wide variety of fruit and vegetables. Fruit and vegetable consumption decreases the risks of coronary heart disease and lung cancer probably to their phenolic compounds [3].

Phenolics are secondary metabolites synthesized by plants, both during normal development and in response to stress conditions such as infection, wounding and UV radiation, among others [4]. In plants, phenolics may act as phytoalexins, antifeedants, attractants for pollinators, contributors to plant pigmentation, antioxidants and protective agents against UV light, among others. In food, phenolics may contribute to their organoleptic properties, such as, bitterness, astringency, color, flavor, odor and oxidative stability [5].

Among colored fruits, containing phenolics there is the Pomegranate [6,7]. The Pomegranate (Punica granatum L. Punicaceae) is a fruit becoming more popular because of its healthy properties [8]. These healthy properties come from the high antioxidant activity of the phenolic compounds content of the Pomegranate fruit [8,9]. However, it is surprising that the literature focused on the antioxidant properties of the phenolics but little is yet known about their contribution to the organoleptic properties of the Pomegranates.

Thus, the present study aims at relating the phenolic contents to the organoleptic properties of the Pomegranate juices as perceived by a panel of participants. To achieve this objective, a collection of seven Tunisian pomegranate cultivars differing in fruit internal color (from light pink to deep pink), taste (from sour-sweet to sweet) and fruit size (from small to large) was examined. The results were used to test for correlations between acceptability tests, physicochemical properties, phenolic contents and antioxidant capacity. This study allows identifying a possible link between the acceptability of Pomegranates and the protective effect of their phenolic compounds.

2. MATERIALS AND METHODS

2.1. Participants

Seventy one adults, eligible and volunteers (40 men and 31 women) were recruited through ad wall in the Institute of Arid Regions of Medenin in Tunisia during the two last weeks of October 2008. Criteria for participation were: age from 18 to 55 years, regular consumer of pomegranate, no experience in sensory analysis, not being allergic to any food and availability for the test. Students and staff of the Institute of Arid Regions who work-ed on pomegranate were excluded in this study as it was thought they may affect the outcomes with their advanced skills on pomegranate taste. Participants reported their answers on individual ballots.

2.2. Samples

Seven Tunisian pomegranate cultivars differing in fruit size, juice color and juice taste were examined (**Table 1**). The fruits were classified as large when their diameter was superior to 95 mm, they were medium if their diameter was from 75 to 95 mm and they were small when their diameter was inferior to 75 mm. The juice color was determined visually while the juice taste was defined according to their sugar to acidity ratio [10]. A sugar to acidity ratio between 8 and 24 indicated sour-sweet cultivars, while a ratio from 25 to 98 referred to sweet cultivars. The cultivars included in the present study were chosen according to customary consumer purchasing practices and the answers to the questionnaire on perceptions and consumption of pomegranate. The denominations of the cultivars refer, in Tunisian, to the population where they belong or to the taste of the fruit [11]. The cultivars were selected from mature fruits grown in the collection of the Tunisian National Germplasm of pomegranate located at Zerkin (33°45'N, 10°16'E) [11]. They were cultivated under homogenous conditions, without any special management (no fertilizers, no irrigation except natural rainfall). Approximately, fifteen kg of each cultivar were picked at harvesting maturity in the last week of September 2008 and in the last week of October 2008. According to Mirdehghan and Rahemi [12], the harvest maturity for pomegranate is achieved, when the arils' weight is greater than that of the peel. Fruits were transported by ventilated car to the laboratory, and trans ferred to a 4°C store room on the same day as they were harvested. To avoid possible contamination of the juices with the metabolites produced by microorganisms, fruits with cracks, cuts, sunburn and other defects in husk were disposed of and only healthy fruits of uniform size and appearance were arranged in one row in wooden boxes containing packing material during the experiments. Whole fruits were preserved for the fruit size acceptability while others were squeezed to juices for further analysis. For thus, fruits for each cultivar were manually peeled, ground in a commercial turmix blender (Moulinex) for 30 s, and filtered through muslin cloth (0.5 mm). The homogenate juices obtained were freshly utilized for physico-chemical analyses and acceptability tests.

Table 1. Codes, cultivars, main characteristics, mean value and standard deviations of instrumental analyses of the seven Tunisian Pomegranate (Punica granatum) studied.

Codes	Cultivars	Main characteristics			TSS (°Brix)	TA (% MA)	TSS/TA	CD	TPC (mgGAE/L)	TFC (mgQE/L)	TAC (mgCGE/L)	CTC (mgCE/L)	ABTS IC_{50} (mg/L)
		Fruit size[A]	Juice color[B]	Juice taste[C]									
CH 4	Chelfi 4	L	LP	Sweet	14.8 ± 0.2	0.2 ± 0.01	59.2 ± 20	0.3 ± 0.01	1073 ± 44	329 ± 15	93 ± 3	2070 ± 0.1	755 ± 21
KL 11	Kalaii 11	L	P	Sweet	14.0 ± 0.1	0.5 ± 0.02	27.8 ± 5	0.5 ± 0.01	1075 ± 30	120 ± 5	188 ± 7	2826 ± 0.1	498 ± 18
GR 2	Garoussi 2	L	DP	Sour-sweet	16.5 ± 0.2	1.8 ± 0.05	9.2 ± 4	1.7 ± 0.04	413 ± 21	268 ± 8	20 ± 1	1436 ± 0.1	1073 ± 22
ZG 11	Zaghouani 11	M	P	Sweet	14.5 ± 0.1	0.4 ± 0.01	36.2 ± 10	1.5 ± 0.03	1316 ± 41	679 ± 21	40 ± 2	2507 ± 0.1	638 ± 21
JB 1	Jbeli 1	M	DP	Sweet	15.8 ± 0.2	0.4 ± 0.01	40.0 ± 20	0.8 ± 0.01	1382 ± 62	1100 ± 32	77 ± 3	2217 ± 0.1	492 ± 17
CH 2	Chelfi 2	S	LP	Sweet	15.0 ± 0.2	0.5 ± 0.02	30.0 ± 10	1.0 ± 0.02	1232 ± 31	264 ± 7	25 ± 1	1458 ± 0.1	1342 ± 33
NB 1	Nabli 1	S	P	Sweet	15.0 ± 0.2	0.4 ± 0.01	37.5 ± 20	0.6 ± 0.01	4570 ± 185	211 ± 8	433 ± 23	5299 ± 0.2	493 ± 15
Mean					15.1 ± 0.2	0.6 ± 0.02	34.3 ± 13	0.9 ± 0.02	1580 ± 59	424 ± 14	125 ± 6	2545 ± 0.1	756 ± 21
Relevant SD					0.8	0.5	15.0	0.5	1357	346	148	1317	333
CV (%)					6	90	44	57	86	82	118	52	44

All values are reported as ± standard deviation; [A]Fruit sizes were based on diameter measurements. Diameter > 95 mm = Large fruit (L), 75 mm < diameter < 95 mm = Medium fruit (M), diameter < 75 mm = Small fruit (S); [B]Juice color was visually determined. LP = Light Pink, P = Pink, DP = Dark Pink; [C]Taste was determined according to Melgarejo, Salazar & Artés, (2000). 8 < TSS/TA < 24 = sour-sweet cultivars; 25 < TSS/TA < 98 = sweet cultivars; TSS = Total Soluble Sugars expressed as degrees Brix (°Brix). TA = Titratable Acidity expressed as pourcent of malic acid (% MA); TSS/TA = sugars to acidity ratio; CD = Color Density; TPC = Total Phenolic Content expressed as mg gallic acid equivalent per liter of juice (mg GAE/L); TFC = Total Flavonoids Content expressed as mg quercetin equivalent per liter of juice (mg QE/L); TAC= Total Anthocyanins Content expressed as mg cyaniding-3-glucoside equivalent per liter of juice (mg CGE/L); CTC = Condensed Tannins Content expressed as mg catechin equivalent per liter of juice (mg CE/L); ABTS IC_{50} = radical scavenging capacity on ABTS radical expressed as concentration of the extract required to reduce 50% of the initial ABTS free radical; Relevant SD = Relevant standard deviation; CV = coefficient of Variation = Relevant SD to mean in percent.

2.3. Sensory Analyses

2.3.1. Questionnaire

Participants were asked to complete a one-page questionnaire (**Table 2**) developed and administered by the author, on their consumption of Pomegranate. The questionnaire included three types of questions: 1) eat, 2) purchase and 3) medicinal effects. For each question, participants indicated their responses by a cross in front of the answer(s) corresponding to their choice. Each participant indicated his answer(s) on an individual questionnaire.

2.3.2. Acceptability Tests

Prior to the acceptability tests, participants were given a presentation outlining the methodology and the procedure of the sensory methods.

Participants evaluated the samples in one session (30 minutes). They were asked to assess, in a first time, the acceptability of the fruit size and the juice color, and, in a second time, to assess the acceptability of sweetness, bitterness and juice overall acceptability. The assessments were done by using a nine-point hedonic scale, where 1 = dislike extremely, 2 = dislike very much, 3 = dislike moderately, 4 = dislike slightly, 5 = neither like nor dislike, 6 = like slightly, 7 = like moderately, 8 = like very much and 9 = like extremely [13]. The samples were presented with 3-digit codes in a balanced order to account for first order and carry-over effects [14]. The whole washed fruits were placed on white plates and the juices were presented in transparent glasses at room temperature. The assessments were done in a well ventilated and at ambient temperature room of the Institute of Arid Regions. Participants were provided with mineral water, unsalted crackers, and expectorant cups to cleanse the palate between sample evaluations. They were instructed to pause for one min between samples.

2.4. Instrumental Analyses

All chemicals used were of analytical reagent grade. All reagents were purchased from Sigma-Aldrich-Fluka (SaintQuentin France).

2.4.1. Physico-Chemical Properties

Titratable Acidity (TA) was determined by potentiometry, using a pH meter, with a 0,1N NaOH solution to a pH of 8.1. The results were expressed as percentage of Malic Acid (%MA). A hand refractometer type OPTECH K7 1319 (Optical Technology, Munich, Germany) graduate of 0° - 32° Brix was used to determine the Total Soluble Solids (TSS). Results were expressed in degrees Brix (°Brix). The sugar to acidity ratio (TSS/TA) was calculated using the relation between Total Soluble Solids by Acidity. Color Density (CD) was determined by measuring the absorbance at 420, 520 and 620 nm in a cell of 1 mm using a spectrophotometer Shimadzu (Shimadzu Corporation, Kyoto, Japan) UV-1201.

Table 2. Questionnaire on perceptions and consumption of pomegranate (N = 71).

Questions on perception and consumption of pomegranate	Frequency	% Responding
1. How often do you eat pomegranate?		
a. Once a day	14	20
b. Every two days	21	29
c. Once a week	11	15
d. Twice a week	26	36
2. What is the most important thing when purchasing pomegranate?		
a. Fruit size	15	21
b. Fruit color	23	33
c. Aril color	11	15
d. Hand-evaluated texture	16	22
e. Cultivar	6	9
3. How do you eat pomegranate?		
a. Salads	6	8
b. Desserts	17	24
c. Fresh fruit	30	42
d. Juice	18	26
4. What is the most important thing when eating pomegranate?		
a. Aril color	19	27
b. Taste	45	63
c. Texture	1	2
d. Nutrition	6	8
5. Would the pomegranate have medicinal effects?		
a. No	22	31
b. Yes	33	46
c. Maybe	16	23
6. If yes or maybe, pomegranate would be:		
a. Anti-carcinogenic	5	7
b. Anti-inflammatory	61	86
c. Anti-alzheimer	1	2
d. Anti-atherosclerotic	4	5

2.4.2. Phenolics Content

The phenolics of each fruit extract were determined by the method of Georgé et al. [15]. The diluted aqueous solution of each extract (0.5 mL) was mixed with Folin Ciocalteu reagent (0.2 N, 2.5 mL). The mixture was allowed to stand at room temperature for 5 min before adding sodium carbonate solution (75 g/L in water, 2 mL). An hour later, the absorbances were measured at 765 nm against a water blank. A standard calibration curve was plotted using gallic acid (0 - 300 mg/L). The results were expressed as mg of gallic acid equivalents (GAE)/L of juice.

1) Flavonoids Content The flavonoids were estimated according to the Dowd method as adapted by Arvouet-Grand, et al. [16]. A diluted

methanolic solution (4 mL) of each plant extract was mixed with a solution (4 mL) of aluminium trichloride ($AlCl_3$) in methanol (2%). The absorbance was read at 415 nm after 15 min against a blank sample consisting of methanol (4 mL) and plant extract (4 mL) without $AlCl_3$. Quercetin was used as reference compound to produce the standard curve. The results were expressed as mg of quercetin equivalents (QE)/L of juice.

2) Anthocyanins Content Total anthocyanin contents were measured with the pH differential absorbance method, as described by Cheng and Breen [17]. Briefly, absorbance of the extract was measured at 510 and 700 nm in buffers at pH 1.0 (hydrochloric acid-potassium chloride, 0.2 M) and 4.5 (acetic acid-sodium acetate, 1 M). The wavelength reading was performed after 15 min of incubation. Anthocyanin contents were calculated using a molar extinction coefficient (ε) of 29600 (cyanidin-3-glucoside) and absorbance of $A = [(A_{510} - A_{700})_{pH\ 1.0} - (A_{510} - A_{700})_{pH\ 4.5}]$. Results were expressed as mg cyanidin-3-glucoside equivalents/ L of juice.

3) Tannins Content Proanthocyanidins reactive to vanillin were analyzed by the vanillin method [18]. One milliliter of extract solution was placed in a test tube together with 2 mL of vanillin (1% in 7 M H_2SO_4) in an ice bath and then incubated at 25°C. After 15 min, the absorbance of the solution was read at 500 nm. Concentrations were calculated as (+)-catechin (mg/L of juice) from a calibration curve.

2.4.3. Antioxidant Capacity

Antioxidant capacity was measured using the improved ABTS method. The radical scavenging capacity of antioxidants for the ABTS (2, 2'-azinobis-3-ethylbenzothiazoline-6-sulphonate) radical cation was determined as described by Re et al. [19]. ABTS radical was generated by mixing a 7 mM of ABTS at pH 7.4 (5 mM NaH_2PO_4, 5 mM Na_2HPO_4 and 154 mM NaCl) with 2.5 mM potassium persulfate (final concentration) followed by storage in the dark at room temperature for 16 h before use. The mixture was diluted with ethanol to give an absorbance of 0.70 ± 0.02 units at 734 nm using

Helios Alpha spectrophotometer (Thermospectronic, USA). For each sample, the diluted methanol solution of the extract (100 μL) was allowed to react with fresh ABTS solution (900 μL), and then the absorbance was measured 6 minutes after initial mixing. Ascorbic acid was used as a standard and the capacity of free radical scavenging was expressed by IC_{50} (mg/L). Values calculated denote the concentration required to scavenge 50% of ABTS radical.

2.5. Statistical Analysis

All measurements were expressed as mean ± standard deviation of triplicate measurements. Standard deviations (SD) did not exceed 5% for the majority of the values obtained. Acceptability data was collected manually on a MS Excel spreadsheet and was recorded with "dislike extremely" equal to one, "like extremely" equal to nine, and intermediate points numbered appropriately. Analysis of Variance (ANOVA) was performed on the qualitative data obtained from the acceptability tests with α set at 0.05. Friedman's test [20], employing a significance level of $\alpha < 0.05$, was applied to the results of the hedonic test, to separate the means of the acceptability tests. Standardized Principal Component Analysis (PCA) was performed on the mean ratings among the participants for all acceptability tests. Hierarchical cluster analysis (HCA) with the Ward criteria was applied on the first two principal components of the PCA [21]. To compare the variability in quantitative physico-chemical composition of the samples having different averages and expressing in different units, the coefficient of variation (CV) was used. It expresses the standard deviation as a percentage of the mean. The data will be considered homogeneous if the CV is less than 15%, and conversely, the data will be considered heterogeneous if the CV is greater than 15%.

A Pearson correlation analysis was carried out between the instrumental measurements with a 95% significance level. A Partial Least Square (PLS) regression analysis was performed to relate the acceptability data (Y-variables) to the instrumental data (X-variables). Regression coefficients for the relations between variables revealed by the PLS regression were estimated by jack-knifing, and significance levels were determined. All the data were collected in spread sheets and the statistical analyses were done using MS Excel Xlstat 2009 (Addinsoft, New York).

3. RESULTS AND DISCUSSION

3.1. Questionnaire

The participants were asked to complete a questionnaire (**Table 2**) about their perceptions and consumptions of pomegranate. The external sensory attributes, such as appearance, color, size and hand-evaluated texture, which are evaluated by the consumer prior to consumption, are intrinsic quality cues. Taste and oral texture evaluated at the time of consumption are experience-quality attributes.

Among intrinsic quality cues, appearance (fruit size and fruit color) was the major factor (one of two) in the purchase of pomegranate fruits for the participants. Of all appearance factors, fruit color (one of three) was suggested to be the most influential quality factor, as participants have expectations of overall quality based on color, while fruit size (one of five) was considered to have minor influence on the consumer evaluation of Pomegranate quality. To our knowledge, no study was conducted on the intrinsic quality cues of pomegranates. Nevertheless, according to Kays [1], color was suggested to be the most influential quality factor, as participants have expectations of overall quality based on color, such as color cues for banana ripeness. The same author notes that at times, color expectations of quality may not be valid because, for example, some orange (Citrus spp.) cultivars are at their optimum when they are green, not orange as most participants perceive.

When experience-quality attributes, participants may be intending to consume pomegranate because of its beneficial health consequences (one of ten), but taste (two of three) and aril color (two of seven) were the fundamental quality that must be satisfied for continued consumption. According to Kays [1], taste is ranked more highly than texture and appearance as a contributor to overall liking for food products; however, color and texture of horticultural products are more frequently cited as consumer quality attributes. Also, pomegranates were preferred as fresh fruit (three of seven) and juice (one of four). These results confirm those of Marlette [22], who found that apple sauce was preferred to whole apples. Thus, fruits would be preferred if they required no preparation, were ready-to-eat or not too difficult to eat.

Results from this study show that the size of the fruit, the taste, the color of the arils and the consumption of the Pomegranate as fresh fruit were the main criteria for participants to select and consummate pomegranate. Thus, these criteria were considered to select the plant material. In order to homogeneity experimental conditions, pomegranate juices were used also for the acceptability tests and the instrumental analyses.

3.2. Acceptability Tests

As seen in **Table 3**, significant differences ($p < 0.05$) existed among pomegranate juices for all the acceptability tests. The participants liked most of the pomegranate juices (as indicated by mean scores > 5) and the average scores for the pomegranate juices were particularly high. Also, participants preferred the juices from the cultivars CH 2 and NB 1 and the cultivar JB 1 received the highest score for juice color acceptability.

Table 3. Mean scores (n = 71) for acceptability tests of the seven Tunisian Pomegranate (Punica granatum) cultivars studied[A].

Cultivars	Fruit size	Juice color	Sweetness	Bitterness	Juice overall acceptability
CH 2	6.42^{b}	5.98^{a}	7.94^{ab}	2.06^{a}	7.20^{ab}
CH 4	7.85^{b}	5.23^{a}	4.57^{ab}	4.01^{a}	6.20^{ab}
GR 2	7.02^{b}	5.14^{a}	3.10^{ab}	5.29^{a}	4.75^{ab}
JB1	5.04^{b}	7.82^{a}	4.00^{ab}	3.57^{a}	5.00^{ab}
KL 11	8.14^{b}	1.34^{a}	4.71^{ab}	4.17^{a}	5.83^{ab}
NB1	4.81^{b}	4.02^{a}	6.29^{ab}	1.43^{a}	6.71^{ab}
ZG 11	5.76^{b}	3.94^{a}	4.96^{ab}	4.63^{a}	4.56^{ab}

[A]Scores were based on a nine-point hedonic scale with 1 = dislike extremely; 5 = neither like nor dislike; and 9 = like extremely. [a-b]Means with different letters within each line indicate significant differences at $p < 0.05$ (Friedman test).

To determine the relationships between the juice overall acceptability of the seven cultivars and the acceptability of the sweetness, the bitterness, the juice color and the fruit size, a PCA was performed. Principal Components 1 (PC1) and 2 (PC2) accounted for 55% and 30% of the total variance, respectively. As shown in **Figure 1**, it seems that participants have clearly distinguished cultivars from each other, since the products were dispersed in the space of the representation. To improve the interpretation of the PCA map, a HCA on the PC1 and PC2 was performed. Three classes of cultivars were revealed (**Figure 2**). A first class, named liked, includes the cultivars CH 2 and NB 1. A second class corresponds to the neither liked nor disliked cultivars, such as, CH 4 and KL 11 and a third class, namely, disliked, was composed of the cultivars GR 2, JB 1 and ZG 11. By representing the three classes of cultivars on the PCA biplot, the liked class was closest to the juice overall acceptability and the acceptability of the sweetness. The neither liked nor disliked and the disliked classes seem to be grouped with the bitterness and diametrically opposed to the juice overall acceptability.

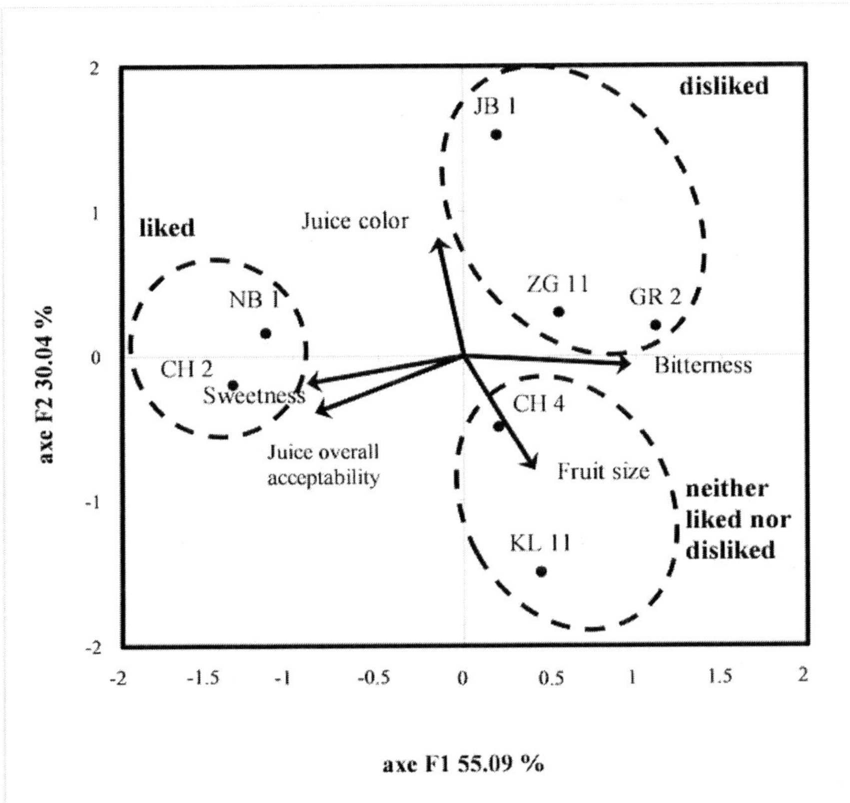

Figure 1. PCA map of consumer acceptability of the seven Tunisian Pomegranate (Punica granatum) cultivars studied. Forms dashed represent the three classes of cultivars defined by the HCA.

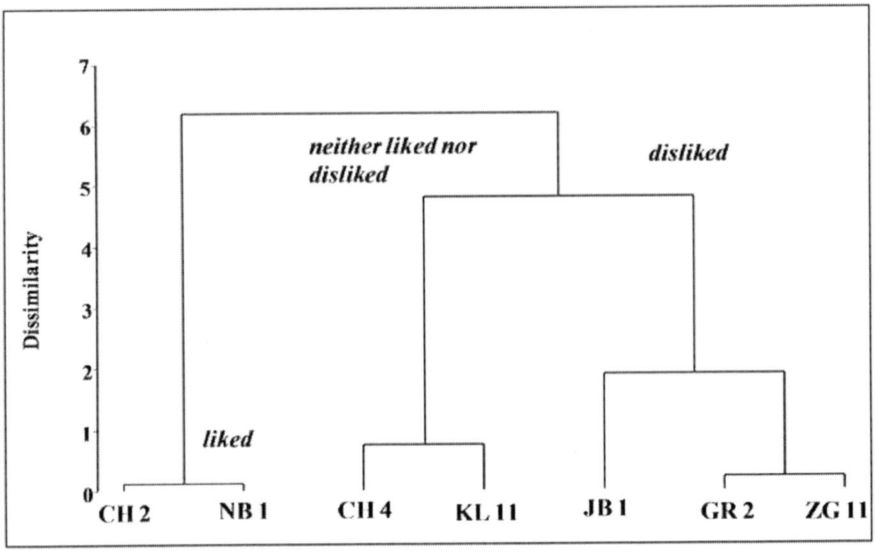

Figure 2. Dendrogram obtained by HCA on the first two principal components of the PCA.

Although, the significant negative correlation ($R = -0.82$, $p < 0.05$) found between the sweetness and the bitterness acceptability explains the natural opposition between the two attributes. Also, the significant correlations found between the juice overall acceptability and the bitterness acceptability on one side ($R = -0.83$, $p < 0.05$), and the sweetness acceptability ($R = 0.82$, $p < 0.05$), on the other, confirmed that heightened perception of bitterness was the principal reason for food rejection [23]. Other significant differences, less expected, were found. In fact, Friedman's test indicated highly significant differences ($p < 0.05$) between fruit size and juice color acceptability, in a hand, and fruit size and bitterness acceptability, in another hand. There is no report, to our knowledge, on the consumer acceptability of Pomegranate.

Thus, it can be concluded that the sweetness accounted for most of the variation in overall acceptability and the consumer rejection of the pomegranate juices was mainly related to the bitterness rather than to the color of juice or the fruit size.

3.3. Physico-Chemical Properties

Table 1 shows the average value and the standard deviations for the physico-chemical properties of the seven pomegranate juices analyzed. As seen, the pomegranate juices differed little in TSS (CV = 6%) with values ranging from 14.0 °Brix for the cultivar CH 4 to 16.5 °Brix for the cultivar GR 2. However, great differences (CV = 90%) were observed in TA with values ranging from 0.2% MA for the cultivar CH 2% to 1.8% MA for the cultivar GR 2. The TSS/TA ratio gives an indication of the sweetness of fruits. The higher the TSS/TA ratio, the sweeter is the fruit. With the highest values for TA and TSS, GR 2 was the sourness cultivar (TSS/TA = 9.2) while with lowest values for TA and TSS, CH 2 was the sweetness cultivar (TSS/TA = 59.2). The TSS of the studied cultivars was close to 15.1 °Brix, both sweet and sour-sweet, thus, the TA was the factor which determines the taste of the Pomegranate. This result was confirmed by the significant negative correlation ($R = -0.85$, $p < 0.05$) found between the TSS/TA ratio and the TA. So, it can be concluded that the taste of the pomegranate juices is inversely related to their TA. Similar conclusion was advanced for orange [24]. **Table 1** shows also that there were great differences (CV = 57%) with respect to the instrumental measurement of color of the pomegranate juices. Juice from the cultivar CH 2 was differentiated as the lighter sample (CD = 0.3) while juice from the cultivar GR 2 was the darkest (CD = 1.7) one; NB 1 cultivar presented intermediate values of lightness (CD = 1.0). The higher CD of the pomegranate juices could be explained by the heat units accumulated (difference between the daily average temperature and 25°C) prior to harvest. Indeed, the intensity of the red color of fresh Israeli pomegranate arils was found to be inversely related to the sum of heat units accumulated during fruit development and ripening [25].

3.4. Phenolic Contents

As shown in **Table 1**, the Total Polyphenols Content (TPC) varied greatly (CV = 86%) among the seven cultivars, and the juice of the cultivar ZG 11 was characterized by the highest TPC with 4570 mg GAE/L, while the lowest TPC was found in the juice of the cultivar GR 2 with 413 mg GAE/L. Özgen, et al. [6] reported a TPC ranging between 1245 and 2076 mg GAE/L for six pomegranate varieties grown in the Mediterranean region of Turkey and Sepúlveda et al. [7] found a TPC from 676 to 1236 mg GAE/L for ten different genotypes of Chilean pomegranate juices. Indeed, phenolic levels increase during exposure to biotic and abiotic stresses, such as wounding, drought, metal toxicity and nutrient deprivation and the highest values of phenolics corresponded to the varieties most sensitive to drought stress [4]. Hence, it can be deduced that the cultivar ZG 11 was the most sensitive to drought stress. This cultivar could be a good source of phenolics.

One of the important quality factors of pomegranate marketing is color. Dietary flavonoids are polyphenolic compounds responsible for the orange, red and blue colors in fruits and vegetables [2]. Also, anthocyanins are a largest group of flavonoids known to intervene in the color of the pomegranate juices [9]. So, Total Flavonoid Content (TFC) and Total Anthocyanin Content (TAC) were measured (**Table 1**). Besides, the cultivar KL 11 exhibited the highest TFC (1100 mg QE/L) while the highest TAC was found in the cultivar ZG 11 (433 mg CGE/L). When consumed regularly, these juices may contribute a significant amount of flavonoid to the diet. Ames [3] suggested that the consumption of flavonoidrich foods protects against human diseases associated with oxidative stress, like cancer. Also, variability among cultivars was greatest for the TFC (CV = 82%) and TAC (CV = 118%). However, the TAC found in this study (20 - 433 mg CGE/L), was three to ten times greater than this reported by Özgen, et al. [6] (6 - 41 mg CGE/L) for six Turkey pomegranate juices. On the other hand, Sepú- lveda et al. [7] founded a TAC around 676 to 1236 mg CGE/L for ten Chilean pomegranate juices. In fact, the variability in the TAC could be explained by the climatic conditions. The extreme temperatures may decrease anthocyanin content by accelerating degradation [26].

Although, Tibe et al. [27] described a relationship between condensed tannin content and color. Thus, the Condensed Tannins Content (CTC) was determined. The pomegranate juices studied exhibited a CV of 52% and the cultivar ZG 11 was characterized by the highest CTC (5299 mg CE /L). Gil et al. [9] found a CTC of 417 mg/L for juice of "Wonderful" pomegranate variety of California while Mousavinejad, et al. [28] reported that the "Sweet Aalak" Iranian cultivar showed a tannin contents of 320 mg/L juice in comparison to the seven other cultivars studied. However, in other species, such as cowpea, Tibe, et al. [27] found that brown colored seeds of cowpeas contain more tannins than cream colored seeds. Therefore, it is not surprising that, the cultivar GR 2 with a red juice color had less tannin than other light colored cultivars in this study. It was unexpected; however, that the cultivar JB 1, with red juice color had a relatively high CTC which may be specific to this cultivar.

When a Pearson correlation was carried out with individual instrumental measurements, a high significant positive correlation ($R = 0.91$, $p < 0.05$) was found between the TPC and the TAC; this suggests that the pomegranates studied are anthocyanin-rich fruits. Another high significant positive correlation ($R = 0.97$, $p < 0.05$) was found between TAC and CTC. This result supports earlier studies [29] that found that anthocyanins and condensed tannins are related through a common biochemical pathway. So, it is important to consider the TAC and the CTC to balance pomegranate juice color.

By studying antioxidant activity of pomegranate juice and its relationship with phenolic composition and processing, Gil et al. [9] concluded that the variations in the phenolic composition may be due to differences among varieties and also, to maturity, horticultural practices, geographic origin, growing season, postharvest storage conditions and processing procedures. To deepen the knowledge on the variation of the phenolic composition of pomegranate juice, the climatic and edaphic influences were limited, as well as, the horticultural practices by collecting the pomegranates from the same site where the culture conditions were homogenous, without fertilizers and irrigation. Also, the postharvest storage conditions of the pomegranate and the processing procedures for the determination of the phenolic composition were the same. So, it can be concluded that the observed differences in terms of phenolic contents of the juices depend on cultivar type or/and ripening season.

3.5. Antioxidant Capacity

The results from free radical scavenging revealed that pomegranate juices differed (CV = 44%) in their antioxidant capacities. Relatively high levels of TPC were observed in ZG 11 and KL 11 cultivars, so they decreased capacity toward ABTS radical scavenging. The hierarchy for antioxidant capacity was KL 11 > ZG 11 > CH 4 > JB 1 > CH 2 > GR 2 > NB 1. Also, antioxidant capacity of the cultivar KL 11 was about three times higher than that of the cultivar NB 1. An antioxidant, which can quench reactive free radicals, can prevent the oxidation of other molecules and may have, therefore, health-promoting effects in the prevention against diseases linked to oxidative stress. Thus, the juices of KL 11 and ZG 11 cultivars may be considered as a source of natural antioxidants for drug products because of their high antioxidant capacity. These results confirm previous works [6, 9,30] about the antioxidant activity of pomegranate juices.

In the present study, phenolic contents were not significantly correlated with antioxidant activity in the ABTS assay. However, the results of exhaustive work of two groups [9,30] on antioxidant activity of pomegranate juices suggested that anthocyanins play only a minor role in this activity. Thus, it can be concluded that the flavonoids, including anthocyanins, and condensed tannins play a minor role in the antioxidant activity of the pomegranate juices studied. Nevertheless, Seeram et al. [8] found that synergism between the antioxidants in a mixture makes the antioxidant activity not only dependent on the concentration, but also on the structure and the interaction between the antioxidants.

3.6. Correlation between Consumer Acceptability Tests and Instrumental Analyses

The PLS regression analysis carried out to relate the five acceptability tests (variables Y) to the nine instrumental analyses (variable X), leads to the map shown in **Figure 3**. With a global R^2 between Y and (t_1, t_2) equal to 0.45, the quality of the regression was low. Also, **Figure 3** shows that fruit size and juice color acceptability cannot be related to pomegranate characteristics since they are located in the centre of the graphical display. Therefore, it does not seem useful to take this acceptability into account. Similarly, it seems that TFC doesn't play an important role in explaining the consumer acceptability of the pomegranate. Therefore, these variables were removed from the analysis. In this approach, a new PLS regression was carried out. The quality of the new PLS regression was noticeable higher than the previous one ($R^2 = 0.71$). By comparing the new map (figure not shown) obtained with the previous one, it can be noticed that the new map is an extract of the **Figure 3**. This confirms the stability and the robustness of the PLS regression analysis.

As shown in **Figure 4**, the sweetness acceptability appeared significantly positively correlated to the CD. Also, the acceptability of bitterness was significantly positively correlated to the TAC. In fact, Zampini, et al. [31], found

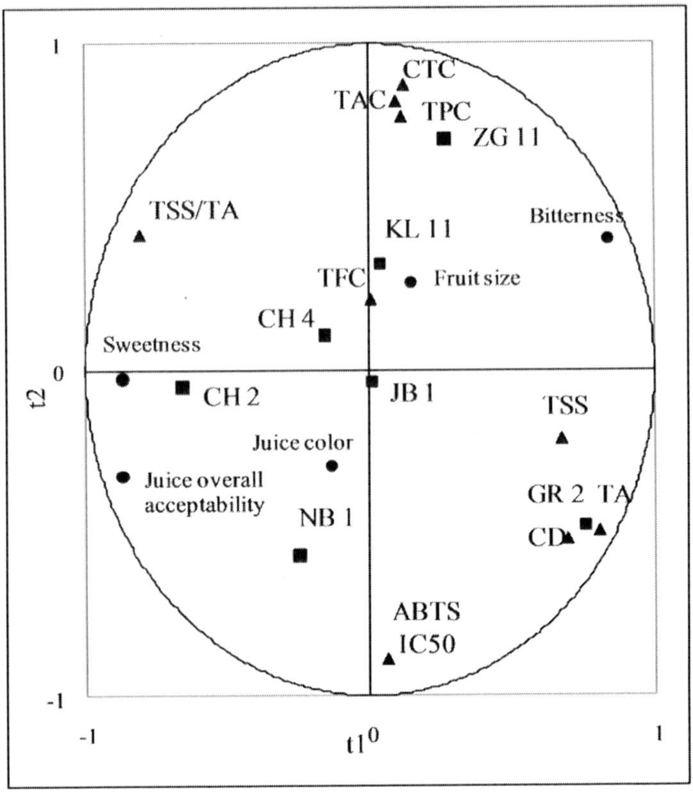

Figure 3. Partial Least Squares regression map showing the correlation between consumer acceptability data and instrumental data in the seven Pomegranates studied (■ = cultivars as indicated in **Table 2**; ● = acceptability attributes; ▲ = instrumental variables; TSS = Total Soluble Sugars expressed as degrees Brix (°Brix); TA = Titratable Acidity expressed as pourcent of malic acid (% MA); TSS/TA = sugars to acidity ratio; CD = Color Density; TPC = Total Phenolic Content expressed as mg gallic acid equivalent per liter of juice (mg GAE/L); TFC = Total Flavonoids Content expressed as mg quercetin equivalent per liter of juice (mg QE/L); TAC = Total Anthocyanins Content expressed as mg cyaniding-3-glucoside equivalent per liter of juice (mg CGE/L); CTC = Condensed Tannins Content expressed as mg catechin equivalent per liter of juice (mg CE/L); ABTS IC_{50} = radical scavenging capacity on ABTS radical expressed as concentration of the extract required to reduce 50% of the initial ABTS free radical).

a significant main effect of flavor on color. Indeed, they revealed that orange-colored solutions were rated as sweeter when they had an orange flavor than when they were flavorless, thus color cues might exert a stronger influence on people's flavor identification responses. Acceptability of bitterness was also significantly negatively correlated to the antioxidant capacities. However, cancer researchers even proposed that heightened bitterness might be a positive feature, allowing consumers to select broccoli sprouts with the highest glucosinolate content [32]. As was also indicated in **Figure 4**, the juice overall acceptability was correlated to the CTC. Previous studies have shown that condensed tannins contribute to the bitterness taste of fruits [23]. The correlations between TAC, CTC and, respectively, bitterness and juice overall acceptability explain the contribution of phenolic compounds to organoleptic properties of the Pomegranates. Nevertheless, according to Gil et al. [9], phenolic compounds are connected to environmental factors, varieties, growing season, postharvest storage conditions and processing procedures. As the cultivars of this study have been grown in the same germplasm, it can be concluded that the organoleptic properties of the Pomegranates studied, like the juice color and the taste were obviously closely connected to the cultivar type or/and ripening season. Further research is needed to investigate which factors may be responsible for the organoleptic properties of the Pomegranates.

According to the current study, it can be concluded that the most appreciated juices were not the most beneficial to health while the fruit size could have a relationship with the beneficial effects of pomegranate on health. To our knowledge, no work relates the consumer acceptability to the antioxidant capacity of the pomegranate. Meanwhile, bitterness in plant foods has been described as a sensory defect with a major economic effect. Responding to taste-driven consumer demand, the food industry generally debittering plant food what improves the taste, but reduces the beneficial effects on health [23]. To avoid these problems, it can be suggested the consumption of cultivars with high

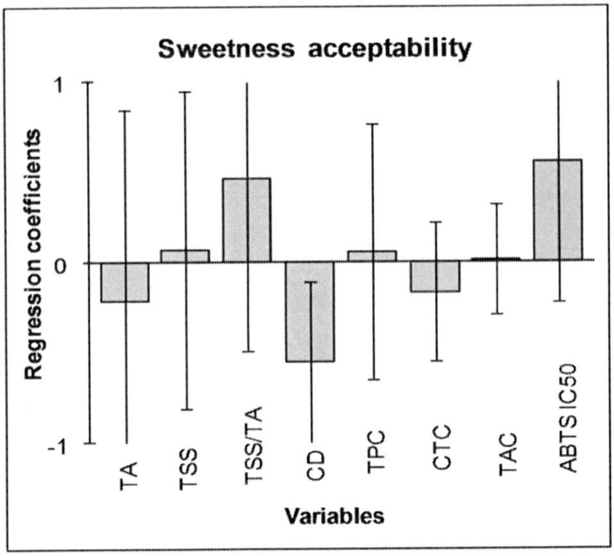
Sweetness acceptability
Regression coefficients
1
0
-1
TA
TSS
TSS/TA
CD
TPC
CTC
TAC
ABTS IC50
Variables

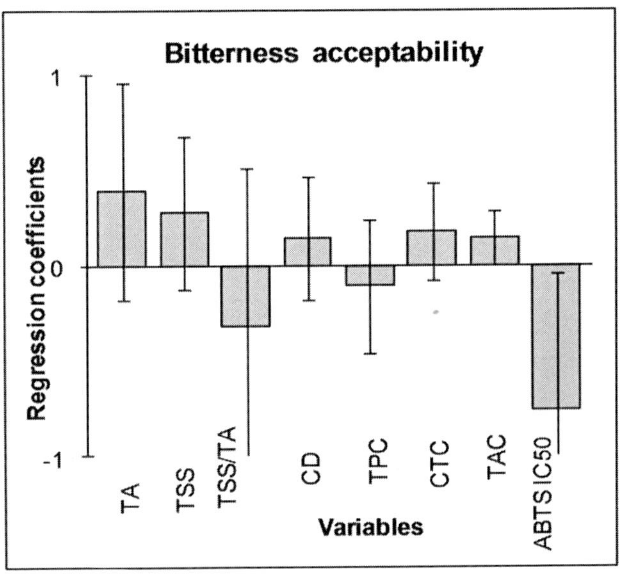
Bitterness acceptability
Regression coefficients
1
0
-1
TA
TSS
TSS/TA
CD
TPC
CTC
TAC
ABTS IC50
Variables

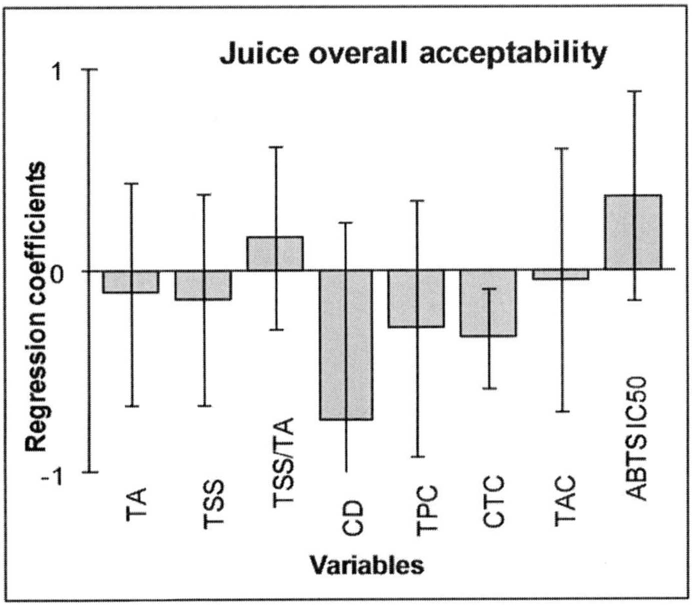

Figure 4. 95% Jack-knife confidence intervals of the Partial Least Squares regression coefficients in the Partial Least Squares regression of sweetness, bitterness and juice overall acceptability on instrumental analyses (Variables with confidence interval includes the value 0 are significant ($p < 0.05$); TSS = Total Soluble Sugars expressed as degrees Brix (°Brix); TA = Titratable Acidity expressed as pourcent of malic acid (% MA); TSS/TA = sugars to acidity ratio; CD = Color Density; TPC = Total Phenolic Content expressed as mg gallic acid equivalent per liter of juice (mg GAE/L); TFC = Total Flavonoids Content expressed as mg quercetin equivalent per liter of juice (mg QE/L); TAC = Total Anthocyanins Content expressed as mg cyaniding-3-glucoside equivalent per liter of juice (mg CGE/L); CTC = Condensed Tannins Content expressed as mg catechin equivalent per liter of juice (mg CE/L); ABTS IC_{50} = radical scavenging capacity on ABTS radical expressed as concentration of the extract required to reduce 50% of the initial ABTS free radical).

concentration of polyphenols, and high antioxidant properties and reduction of the bitterness by addition of sugar, such as in the case of wines. The addition of sugar is a relatively inexpensive option to mask the bitterness while preserving the bioactive potential of the polyphenols.

However, further studies should be undertaken to elucidate the duality product acceptability and health benefits.

4. CONCLUSIONS

The frequency of respondents to the questionnaire showed that three of seven of the participants eat Pomegranate as fresh fruit and that the size of the fruit, the taste and the color of the aril were the main criteria for participants to select Pomegranate. This study also found that the taste of the pomegranate juices was inversely related to their TA. In addition, the juices studied appeared to have a great variability in terms of TPC, CTC, TFC and TAC, as well as, for the antioxidant capacities. PLS regression revealed that the juices most appreciated by participants were not the most beneficial to health. However, a significant correlation was found between the fruit size acceptability and the antioxidant capacity of the Pomegranate, so, a relationship could be established between the size of the Pomegranate and its health benefits.

By limiting the climatic and edaphic influences, as well as, the horticultural practices by collecting the pomegranates from the same site, it can be concluded that the observed differences in terms of phenolic contents and organoleptic properties of the pomegranate juices studied depend on cultivar type or/and ripening season. Although the primary function of food is to provide nutrients, its secondary function concerns organoleptic properties such as taste and its tertiary function is to prevent disease. However, results of this study appeared that the secondary and tertiary functions are linked.

The results of this study could be used to understand the complex nature between the acceptability of a product, as perceived by untrained subjects, and its benefits on health.

5. ACKNOWLEDGEMENTS

The authors are grateful for the subjects who participated in the acceptability tests. They also kindly thank Miss Amina BARKHALLAH for proof reading the English version of this manuscript.

REFERENCES

1. S. J. Kays, "Preharvest Factors Affecting Appearance," Postharvest Biology and Technology, Vol. 15, No. 3, 1999, pp. 233-247.

2. H. M. Merken and G. R. Beecher, "Measurement of Food Flavonoids by High-Performance Liquid Chromatography—A Review," Journal of Agricultural and Food Chemistry, Vol. 48, No. 3, 2000, pp. 577-599.

3. B. N. Ames, "Dietary Carcinogens and Anticarcinogens: Oxygen Radicals and Degenerative Diseases," Science, Vol. 221, No. 4617, 1983, pp. 1256-1263.

4. B. Winkel-Shirley, "It Takes a Garden. How Work on Diverse Plant Species Has Contributed to an Understanding of Flavonoid Metabolism," Journal of Plant Physiology, Vol. 127, No. 4, 2001, pp. 1399-1404.

5. F. Shahidi and M. Naczk, "Phenolics in Food and Nutraceuticals: Sources, Applications and Health Effects," CRC Press, Boca Raton, 2004.

6. M. Özgen, C. K. Durgac, S. Serc and C. Kaya, "Chemical and Antioxidant Properties of Six Pomegranate Cultivars Grown in the Mediterranean Region of Turkey," Food Chemistry, Vol. 111, No. 3, 2008, pp. 703-706.

7. E. Sepúlveda, C. Sáenz, A. Peña, P. Robert, B. Bartolomé and C. Gómez-Cordovés, "Influence of the Genotype on the Anthocyanin Composition, Antioxidant Capacity and Color of Chilean Pomegranate (Punica granatum L.) Juices," Chilean Journal of Agricultural Research, Vol. 70, No. 1, 2010, pp. 50-57.

8. N. Seeram, R. N. Schulman and D. Heber, "Pomegranates: Ancient Roots to Modern Medicine," CRC Press/Taylor and Francis, Boca Raton, 2006.

9. M. I. Gil, F. A. Tomas-Barberan, B. Hess-Pierce, D. M. Holcroft and A. A. Kader, "Antioxidant Activity of Pomegranate Juice and Its Relationship with Phenolic Composition and Processing," Journal Agriculture Food Chemistry, Vol. 48, No. 10, 2000, pp. 4581-4589.

10. P. Melgarejo, D. M. Salazar and F. Artés, "Organic Acids and Sugars Composition of Harvested Pomegranate Fruits," European Food Research and Technology, Vol. 211, No. 3, 2000, pp. 185-190.

11. M. Mars, "Ressources Génétiques du Grenadier (Punica granatum L.) en Tunisie: Propspection, Conservation et Analyse de la Diversité," Thèse de Doctorat es Sciences Naturelles, Faculté des Sciences de Tunis, Université El-Manar, Tunisie, 2001.

12. S. H. Mirdehghan and M. Rahemi, "Seasonal Changes of Mineral Nutrients and Phenolics in Pomegranate (Punica granatum L.) Fruit," Journal of Scientia Horticulturae, Vol. 111, No. 2, 2007, pp. 120-127.

13. M. Meilgaard, G. V. Civille and B. T. Carr, "Sensory Evaluation Techniques," 2nd Edition, CRC Press, Boca Raton, 1991, pp. 354.

14. H. J. H. MacFie, N. Bratchell, K. Greenhoff and L. Vallis, "Designs to Balance the Effect of Order of Presentation and First-Order Carry-Over Effects in Hall Tests," Journal of Sensory Studies, Vol. 4, No. 2, 1989, pp. 129-148.

15. S. Georgé, P. Brat, P. Alter, and M. J. Amiot, "Rapid Determination of Polyphenols and Vitamin C in Plant-Derived Products," Journal Agriculture Food Chemistry, Vol. 53, No. 5, 2005, pp. 1370-1373.

16. A. Arvouet-Grand, B. Vennat, A. Pourrat and P. Legret, "Standardisation d'un Extrait de Propolis et Identification des Principaux Constituants," Journal de Pharmacie de Belgique, Vol. 49, No. 6, 1994, pp. 462-468.

17. G. W. Cheng and P. J. Breen, "Activity of Phenylalanine Ammonialyase (PAL) and Concentrations of Anthocyanins and Phenolics in Developing Strawberry Fruit," American Society for Horticultural Science, Vol. 116, No. 5, 1991, pp. 865-869.

18. M. L. Price, S. Van Scoyoc and L. G. Butler, "A Critical Evaluation of the Vanillin Reaction as an Assay for Tannin in Sorghum Grain," Journal Agriculture Food Chemistry, Vol. 26, No. 5, 1978, pp. 1214-1218.

19. R. Re, N. Pellegrini, A. Proteggente, A. Pannala, M. Yang and C. Rice-Evans, "Antioxidant Activity Applying an Improved ABTS Radical Cation Decolorization Assay," Free Radical Biology and Medicine, Vol. 26, No. 9-10, 1999, pp. 1231-1237.

20. M. O'Mahony, "Sensory Evaluation of Food: Statistical Methods and Procedures," Marcel Dekker, New York, 1986.

21. I. T. Jolliffe, "Principal Component Analysis," 2nd Edition, Springer, 2002, 487 p.

22. M. A. Marlette, S. B. Templeton and M. Panemangalore, "Food Types, Food Preparation, and Competitive Food Purchases Impact School Lunch Plate Waste by SixthGrade Students," Journal of the American Dietetic Association, Vol. 105, No. 11, 2005, pp. 1179-1182.

23. R. L. Rouseff, "Bitterness in Food Products: An Overview," In: R. L. Rouseff, Ed., Bitterness in Foods and Beverages. Developments in Food Science, Elsevier, Amsterdam, Vol. 25. 1990, pp. 1-14.

24. F. Karadeniz, "Main Organic Acid Distribution of Authentic Citrus Juices in Turkey," Turkish Journal of Agriculture and Forestry, Vol. 28, No. 4, 2004, pp. 267-271.

25. H. Borochov-Neori, S. Judeinstein, E. Tripler, M. Harari, A. Greenberg, I. Shomer and D. Holland, "Seasonal and Cultivar Variations in Antioxidant and Sensory Quality of Pomegranate (Punica granatum L.) Fruit," Journal of Food Composition and Analysis, Vol. 22, No. 3, 2009, pp. 189-195.

26. L. Shaked-Sachary, D. Weiss, M. Reuveni, A. NissimLevi and M. Oren-Shamir, "Increased Anthocyanin Accumulation in Aster Flowers at Elevated Temperatures Due to Magnesium Treatment," Plant Physiology, Vol. 114, No. 4, 2002, pp. 559-565.

27. O. Tibe, J. O. Amarteifio and R. M. Njogu, "Trypsin Inhibitor Activity and Condensed Tannin Content in Bambara Groundnut (Vigna subterranean L. verde) Grown in Southern Africa," Journal of Applied Sciences and Environmental Management, Vol. 11, No. 2, 2007, pp. 159- 164.

28. G. Mousavinejad, Z. Emam-Djomeh, K. Rezaei and M. H. H. Khodaparast, "Identification and Quantification of Phenolic

Compounds and Their Effects on Antioxidant Activity in Pomegranate Juices of Eight Iranian Cultivars," Food Chemistry, Vol. 115, No. 4, 2009, pp. 1274-1278.

29. L. J. Porter, L. N. Stich and B. G. Chan, "The Conversion of Procyanidins and Prodelphinidins to Cyaniding and Delphinidin," Phytochemistry, Vol. 25, No. 1, 1986, pp. 223-230.

30. R. Tzulker, I. Glazer, I. Bar-Ilan, D. Holland, M. Aviram and R. Amir, "Antioxidant Activity, Polyphenol Content and Related Compounds in Different Fruit Juices and Homogenates Prepared from 29 Different Pomegranate Accessions," Journal of Agricultural and Food Chemistry, Vol. 55, No. 23, 2007, pp. 9559-9570.

31. M. Zampini, E. Wantling, N. Phillips and C. Spence, "Multisensory Flavor Perception: Assessing the Influence of Fruit Acids and Color Cues on the Perception of FruitFlavored Beverages," Food Quality and Preference, Vol. 19, No. 3, 2008, pp. 335-343.

32. J. W. Fahey, Y. Zhang and P. Talalay, "Broccoli Sprouts: An Exceptionally Rich Source of Inducers of Enzymes That Protect against Chemical Carcinogens," Proceedings of the National Academy of Sciences of the United States of America, Vol. 94, No. 19, 1997, pp. 10367-10372.

6

CHAPTER

Comparison of Fresh and Commercial Pomegranate Juices fromMollar de Elche Cultivar Grown under Conventional or Organic Farming Practices

Nallely Nuncio-Jáuregui [1], Marina Cano-Lamadrid [1], Francisca Hernández [2], Ángel A. Carbonell-Barrachina [1] and Ángel Calín-Sánchez [1,*]

[1]Research Group in Food Quality and Safety, Food Technology Department, University Miguel Hernandez, Ctra. Beniel, km 3.2, 03312-Orihuela, Alicante, Spain

[2]Research Group in Plant Production and Technology, Plant Science and Microbiology Department, University Miguel Hernandez, Ctra. Beniel, km 3.2, 03312-Orihuela, Alicante, Spain

ABSTRACT

Pomegranate juice has gained a high reputation for its health properties and consequently is now a highly demanded product. However, there is an interest in knowing if there are differences between conventional and organic agricultural practices. For that reason, total phenolic content, antioxidant capacity, punicalagin isomers and sensory analysis of pomegranate juices from *Mollar de Elche* cultivar were studied. A comparison between fresh and commercial pomegranate juice obtained by conventional and organic agricultural practices was conducted. The total phenolic content values ranged from 2285 to 2457 mg GAE L^{-1}; however, there was no significant difference among conventional and organic juices. The antioxidant capacity evaluated by DPPH and ABTS methods showed significant differences ($p < 0.05$) among conventional and organic juices. The antioxidant capacity values ranged from 17.7 to 35.9 mmol Trolox L^{-1} for DPPH and from 5.09 to 27.9 mmol Trolox L^{-1} for ABTS. Significant differences ($p < 0.05$) were observed among conventional and organic juices in punicalagin isomer, with the highest value found in conventional fresh pomegranate juice (0.48 g L^{-1}). Descriptive sensory analysis showed that fresh pomegranate, fresh rind, earthy, vegetal, bitter, and astringent notes were higher in conventional fresh pomegranate juice. Cooked and mushroom notes predominated in conventional commercial pomegranate juice; while the organic juice was characterized by fresh pomegranate, fresh rind, earthy and sweet notes.

Keywords: pomegranate juice; phenolic content; antioxidant capacity; sensory analysis

1. INTRODUCTION

Conventional farming systems are experiencing a serious crisis due to the existence of a highly industrialized agriculture, using more intensively chemical and synthetic products [1]. In some cases, conventional farming is being replaced by organic farming, which does not use chemical fertilizers, pesticides and considers as "a holistic view of agriculture that aims to reflect the profound interrelationship that exists between farm biota, its production and the overall environment" [2]. This farming practice is based on rotating crops and using green manure to maintain the productivity of soil without damaging the environment, producing food free of any chemical residue [3]. It is thought that in the absence of pesticides, plants could contain higher levels of antioxidant components as a result of enhanced synthesis of active phytochemicals produced in defense against biotic and abiotic stress [4]. This concern arises because conventional agricultural practices utilize levels of pesticides and fertilizers that can result in a disruption of natural production of phenolic metabolites in the plant [5].

In the last years pomegranate fruit (*Punica granatum* L.) has taken great attention for its health benefits. The edible portion of pomegranates (arils) can be consumed in fresh. However, there is a huge amount of fruits which quality is not good enough to be consumed in fresh, mainly because of physiopathies [6]. Certainly, consumers' acceptance of these low quality fruits would be low. For this reason, it is necessary to find a commercial application for the fruits which cannot be commercialized in fresh; the main option is to use them in the "juice industry".

Numerous studies on the antioxidant capacity and total phenolic content have shown that pomegranate juice contains high levels of antioxidants, higher than most other fruit juice and beverages [7,8,9]. The main antioxidant compounds in pomegranate juice are hydrolysable tannins, but anthocyanins and ellagic acid derivatives also contribute to the total antioxidant capacity of the juice [8]. Therefore, pomegranate juice has been recommended as a preventive treatment for coronary heart disease [10] and to improve the chemotherapeutic effects on human prostate cancer [11].

Pomegranate juice is being marketed as high quality or "gourmet" items because of its recent high popularity and price. The increasing concerns of food consumers on the relationship between health and diet, as well as on environmental issues have created new perspectives to the consumption of organic food [12]. There is controversy over the differences in the quality of conventional and organic pomegranate juices; while some studies [13,14] showed evidences of organic food being more nutritious, others [15,16] concluded that there were no consistent nutritional differences between both types of juices.

Therefore, the aim of the present study was to evaluate and compare the total phenolic content, antioxidant capacity, contents of punicalagin isomers, and sensory analysis of fresh and commercial pomegranate juices from *Mollar de Elche* cultivar grown using both conventional and organic agricultural practices.

2. MATERIALS AND METHODS

2.1. Plant Material and Sample Processing

Four pomegranate juices were used for the analysis: conventional fresh pomegranate juice (c-FPGJ), organic fresh pomegranate juice (o-FPGJ), conventional commercial pomegranate juice (c-CPGJ), and organic commercial pomegranate juice (o-CPGJ). The commercial pomegranate juices were directly acquired from companies in Alicante with which there is previously company-university collaboration. Commercial pomegranate juices in study (VitalGrana) were elaborated with "*Mollar de Elche*" cultivar and have two different presentations: (1) conventional and (2) organic.

Pomegranate fruits, "*Mollar de Elche*" cultivar, were harvest at fruit ripening stage (October 2014). Conventional fruits were obtained from one of the main pomegranate gene banks of the European Union, which is located at the experimental field station of the Miguel Hernandez University in Orihuela, Alicante, Spain (02°03'50"E, 38°03'50"N, and 25 m above sea level). Organic fruits were obtained from organic field certified by CAERM (Council Agriculture the Ecological Region of Murcia) in Alquerias (Murcia), compliance with current legislation: R (CE) 834/2007 and R (CE) 889/2008.

Both types of organic juices were produced using the farming practices approved by the CAERM (*Consejo de Agricultura Ecológica de la Región de*

Murcia, http://www.caermurcia.com/introduction). Both types of juices were marketed with the seal of this official organism.

Ten fruits from each field were harvested in the same state of ripeness and immediately transported to the laboratory. Each husk was carefully cut at the equatorial zone with a sharp knife, and then arils were manually extracted to obtain the fresh pomegranate juice. The following parameters were analyzed in fresh and commercial juices: total phenolic content, contents of punicalagins (α and β), antioxidant capacity and sensory analysis. The chemical analyses were run in triplicate.

2.2. Extraction Procedure for Total Phenolic Content and Antioxidant Capacity (DPPH and ABTS Methods)

For a better extraction and analysis of antioxidant compounds, the method described by Chong *et al.* [17] was used. Briefly, pomegranate juice (2 mL) were mixed with 10 mL of MeOH/water (80:20 v/v) + 1% HCl, and sonicated at 20 °C for 15 min and left for 24 h at 4 °C. Then, the extract was again sonicated for 15 min, and centrifuged at 15,000 rpm for 10 min. Finally, the supernatant was collected and used for the analyses of total phenolic content and antioxidant capacity.

2.3. Total Phenolic Content

Total phenolic content (TPC) was quantified using Folin-Ciocalteu colorimetric method described previously by Gao *et al.* [18]. Juice extracts (0.1 mL) were mixed with 0.2 mL of Folin-Ciocalteu reagent and 2 mL of ultrapure water. Then, the mixture was incubated at room temperature for 3 min and 1 mL of 20% sodium carbonate was added. TPC were determined after 1 h of incubation at room temperature. The absorbance of the resulting blue color was measured at 765 nm using an UV-Vis Uvikon XS spectrophotometer (Bio-Tek Instruments, Saint Quentin Yvelines, France). Quantification was done with respect to the standard curve of gallic acid. The results were expressed as gallic acid equivalents (GAE), mg GAE per liter of juice.

2.4. Antioxidant Capacity

The antioxidant activity (AOC) was evaluated by the DPPH and ABTS methods. The DPPH (radical 2,2-diphenyl-1-picrylhydrazyl) method was using as described by Brand-Williams *et al.* [19] with a modification in the reaction time. Briefly, 10 μL of the supernatant were mixed with 40 μL of MeOH and added to 950 μL of DPPH solution. The mixture was shaken vigorously and placed in a dark room for 10 min. The decrease in absorbance was measured at 515 nm in UV-Vis Uvikon XS spectrophotometer (Bio-Tek Instruments, Saint Quentin Yvelines, France).

The ABTS (2,2-azinobis-(3-ethylbenzothiazoline-6-sulfonic acid)) radical cation was employed according to [19]. The ABTS was dissolved in water to a 7 mM concentration; the radical cation ($ABTS^{\bullet+}$) was produced by reacting ABTS stock solution with 2.45 mM potassium persulfate and kept in the dark at room temperature for 16 h before use. Then, 10 μL of the supernatant were mixed with 990 μL of ABTS. After 10 min of reaction, the absorbance was measured at 734 nm. The absorbance was measured in UV-Vis Uvikon XS spectrophotometer (Bio-Tek Instruments, Saint Quentin Yvelines, France). Calibration curves for DPPH and ABTS methods were prepared with trolox (6-hydroxy-2,5,7,8-tetramethylchroman-2-carboxylic acid) showing good linearity ($R^2 \geq 0.998$). The results were expressed as mean ± standard error and units in mmol Trolox (vitamin E analogous) per liter of juice.

2.5. Identification and Quantification of Punicalagin Isomers

Punicalagins (α and β) were determined in commercial and fresh juice (conventional and organic juices), 5 mL diluted with 5 mL of

MeOH/water (80:20 v/v) and 1% acetic acid and sonicated at 20 °C for 15 min and left for 24 h at 4 °C. Then the extract was again sonicated for 15 min, and centrifuged at 15,000 rpm for 20 min. Supernatants were filtered through a 0.45-μm Millipore filter and then injected into a Hewlett-Packard HPLC series 1200 equipped with a diode-array detector (Agilent, Waldbronn, Germany). Each sample (20 μL) was analyzed on a LiChroCART 100 RP-18 reversed-phased column (250 × 4 mm, particle size 5 μm; Merck, Darmstadt, Germany) equipped with a pre-column C18 (LiChrospher 100 RP-18, 5 μm; Merck, Darmstadt, Germany) using a mobile phase of 1% acetic acid in ultra-high-purity deionized water (solvent A) and 1% acetic acid in MeOH (solvent B). Elution was performed at flow rate of 1 mL/min using a gradient starting with 1% B for 5 min, and increasing to 60% B at 40 min. Punicalagins detection were conducted at 360 nm. For the identification of punicalagins, absorption spectra and retention times were employed and compared with those obtained from the chemical standards. The results were expressed as mean ± standard error and units in g per liter of juice.

2.6. Sensory Evaluation

Sensory evaluation with trained panel was used to compare fresh and commercial pomegranate juices (conventional and organic). A panel of 8 panelists, ages 20 to 55 years (4 female and 4 male) was trained in descriptive evaluation of pomegranate juices. All panelists work at Miguel Hernandez University and have a wide expertise in sensory evaluation of fruits and juices [20].

Panelists were asked to evaluate the intensity of the next attributes: (1) *appearance and aroma*: color, fresh pomegranate, fresh rind, earthy, vegetal, mushrooms and cooked; and (2) *flavor and basic tastes*: fresh pomegranate, fresh rind, earthy, vegetal, mushrooms, sweet, sour, bitter and astringent. Panelists used for the evaluation an 11-point scale, where 0 was extremely low intensity, 5 was regular intensity, and 10 was extremely high intensity.

2.7. Statistical analysis

Data from the analyses of pomegranate commercial and fresh juices (conventional and organic juice) were examined first by analysis of variance for mean comparison. Later, the method used to discriminate among the means (Multiple Range Test) was Tukey's procedure. Significance was defined at $p \leq 0.05$. Statistical analyses were performed using StatGraphics Plus 5.0 software (Manugistics, Inc., Rockville, MD, USA).

3. RESULTS AND DISCUSSION

3.1. Total Phenolic Content

Total phenolic content (TPC) in fresh and commercial (conventional and organic) pomegranate juice, are presented in Table 1. The TPC values ranged from 2285 (o-CPGJ) to 2457 (c-FPGJ) mg GAE L^{-1}. The variability in the TPC values may be caused by reactions taken place during juice manufacture (*i.e.*, hydroxylation, methylation, isoprenylation, imerization and/or glycosylation) [21]. The obtained TPC values were similar with those found by Mena *et al.* [22] (1562 to 2342 mg GAE L^{-1}) in "*Mollar de Elche*" and "*Valenciana*" cultivars. In a similar way, the farming type (organic or conventional) did not affect the TPC in orange juice [23]. Contrary, Dani *et al.* [24] reported higher phenolic content in organic grape juice compared to conventional grape juice. Therefore, the effect of organic or conventional growing conditions is not clear yet. Probably, there is a combination of conditions which influences on the phenolic content such as specie, cultivar, climate conditions, *etc.* The current study was the first one on pomegranate juice phenolic content as affected by organic and conventional agricultural conditions, and not statistically significant differences were found.

Table 1. Total phenolic content and antioxidant capacity in fresh and commercial pomegranate juices obtained from conventional and organic agricultural practices: (1) total phenolic content, TPC (mg GAE L^{-1}), (2) DPPH antioxidant capacity, DPPH-AOC (mmol Trolox L^{-1}) and (3) ABTS antioxidant capacity, ABTS-AOC (mmol Trolox L^{-1}).

Sample	TPC (mg GAE L^{-1})	DPPH-AOC (mmol Trolox L^{-1})	ABTS-AOC
c-FPGJ	2457[†] ± 31[‡]	17.7 ± 0.2 b	7.58 ± 0.61 b
o-FPGJ	2384 ± 7	35.9 ± 0.6 a	27.9 ± 2.3 a
c-CPGJ	2304 ± 10	17.8 ± 0.1 b	5.09 ± 0.39 b
o-CPGJ	2285 ± 12	17.9 ± 0.1 b	11.6 ± 1.2 b
ANOVA	N.S.	***	***

c-FPGJ: conventional fresh pomegranate juice; o-FPGJ: organic fresh pomegranate juice; c-CPGJ: conventional commercial pomegranate juice; o-CPGJ: organic commercial pomegranate juices; ANOVA: analysis of variance. † Values are the mean of 3 replications (±standard error). ‡ Values followed by the same letter within the same column were not statistically different at $p < 0.05$ (according to Tukey's multiple range test). N.S.: not significant ($p < 0.05$) and *** significant at $p < 0.001$.

3.2. Antioxidant Capacity

Table 1 shows the results of antioxidant capacity of fresh and commercial (conventional and organic) pomegranate juices by DPPH and ABTS methods. Significant differences ($p < 0.05$) were observed among conventional and organic juices. The AOC values ranged from 17.7 and 35.9 mmol Trolox L^{-1} for DPPH and from 5.09 to 27.9 mmol Trolox L^{-1} for ABTS. These values are higher in comparison with those found by Gil *et al.* [8] for red wine (6–7.5 mmol Trolox L^{-1}) and green tea (4.4 mmol Trolox L^{-1}). In general the values obtained for all pomegranate juices were to those previously reported by Mena *et al.* [22], which reported values in the range from 7.01 and 37.1 mmol Trolox L^{-1}.

Using the ABTS and DPPH methods, antioxidant capacity of the experimental pomegranate juices showed that fresh juice has higher antioxidant capacity than commercial pomegranate juice, specially the organic juice. Organic farming is supposed to optimize natural resources; however, conventional agriculture used non-natural products, such as fertilizers, and pesticides. Due to this supplementation, conventional farming is thought to accumulate less content of secondary metabolites, which induce the antioxidant capacity and consequently some health benefits since the most natural farming the highest content of bioactive compounds [23,24]. On the other hand, the total antioxidant capacity of pomegranate juice is attributed to polyphenols originating from the peels [8]. If whole fruits are pressed to prepare fresh juices (this is not the case of the current study, in which juices were prepared only using arils), it would be expected a large amount of bioactive compounds from the peels and consequently fresh juices would have high antioxidant capacity. Likewise, processing and pasteurization conditions play an important role in the antioxidant capacity of juice, since the bioactive compounds are affected by extrinsic factors such as oxygen, light and especially temperature [25,26,27].

Up to now, there are not scientific literature regarding the AOC of pomegranate juices from fruits cultivated by organic agricultural practices; however, several studies have been conducted on the TPC of different fruits. For instance, organic juices from apple, pear, blackcurrant, carrot, beetroot and celery showed slightly higher TPC than conventional ones [28]; however, no mention was made about the AOC of these juices.

3.3. Identification and Quantification of Punicalagin Isomers

Table 2 shows the results of punicalagin isomers content in the different pomegranate juices under study. α-Punicalagin content ranged from 0.07 to 0.48 g L^{-1}, while β-punicalagin content ranged from 0.03 to 0.22 g L^{-1}. In general the c-FPGJ juice showed the highest value. The results showed that α-punicalagin was more abundant than β-punicalagin. The most abundant compounds among the pomegranate polyphenols are punicalagin isomers [29], which have important antioxidant and atherosclerotic biological properties [30]. The results from the present study showed significant difference among fresh and commercial pomegranate juice, as well as conventional and organic agricultural practices. These differences may be due to the temperatures used during industrial processing, storage conditions, cultivation techniques, and/or fertilization level [26,31,32].

Table 2. Contents of α-punicalagin and β-punicalagin (g L^{-1}) in fresh and commercial pomegranate juices obtained from conventional and organic agricultural practices.

Sample	α-Punicalagin	β-Punicalagin
	(g L^{-1})	
c-FPGJ	0.48 ± 0.01 a	0.22 ± 0.01 a
o-FPGJ	0.21 ± 0.04 b	0.12 ± 0.01 b
c-CPGJ	0.12 ± 0.01 b	0.08 ± 0.01 c
o-CPGJ	0.07 ± 0.01 c	0.03 ± 0.01 c
ANOVA	***	***

c-FPGJ: conventional fresh pomegranate juice; o-FPGJ: organic fresh pomegranate juice; c-CPGJ: conventional commercial pomegranate juice; o-CPGJ: organic commercial pomegranate juices; ANOVA: analysis of variance. † Values are the mean of 3 replications (±standard error). ‡ Values followed by the same letter within the same column were not statistically different at $p < 0.05$ (according to Tukey's multiple range test). *** significant at $p < 0.001$.

3.4. Sensory Analysis

Regarding appearance and aroma attributes during descriptive sensory analysis, c-FPGJ and o-FPGJ showed higher scores of desirable attributes (color, fresh pomegranate aroma, fresh rind aroma and vegetal notes) for fresh juices than those processed juices (Figure 1A). Similar situations took place when comparing fresh and commercial pomegranate juices in previous study [32].

c-CPGJ showed the highest scores in undesirable attributes, such as cooked and mushrooms notes; these attributes are typical from commercial juices after heat treatment [32]. o-CPGJ obtained lower scores of cooked notes showing closer values to desirable attributes of appearance and aroma of fresh juices.

Taking into account, flavor and basic tastes, c-FPGJ showed higher scores in fresh pomegranate, earthy, bitterness, and astringency (Figure 1B). These results were well correlated with the highest values found in the total phenolic content of c-FPGJ. Phenolic content is related to the astringency and bitterness of the juice, so high polyphenols content can cause unpleasant levels of bitterness and astringency [33]. The o-FPGJ was characterized by fresh pomegranate, fresh rind and earthy notes. The c-CPGJ has the highest values in mushrooms, cooked and sour notes. Finally, the o-CPGJ showed the highest values in sweet notes.

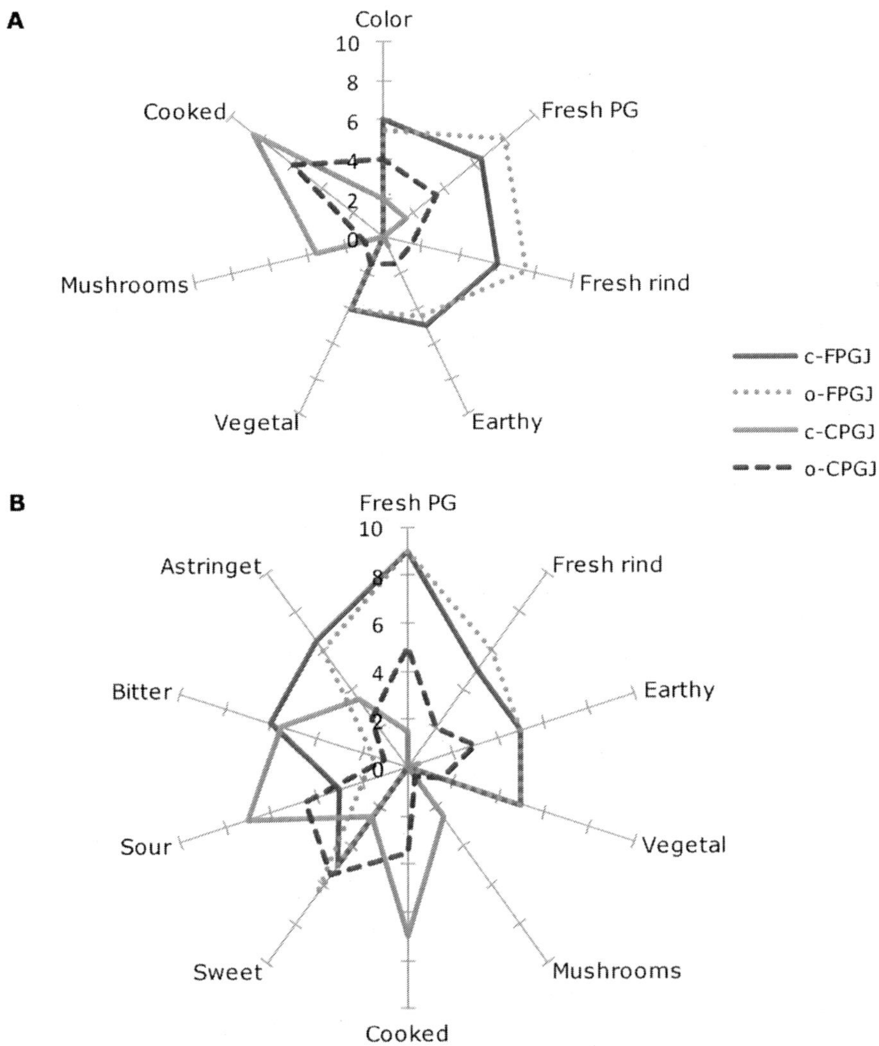

Figure 1. Descriptive Sensory Analysis by a trained panel of fresh and commercial pomegranate juices obtained from conventional and organic agricultural practices (A = appearance and aroma; B = flavor and basic tastes).

4. CONCLUSIONS

This first study comparing fresh and commercial pomegranate (cultivar *Mollar de Elche*) juices from organic and conventional agricultural practices provided useful information about the antioxidant capacity, total phenolic content, punicalagina isomers content and sensory profiles of these samples. Total phenolic content did not show significant differences in all the juices under study, while antioxidant capacity was significantly higher for organic fresh pomegranate juice. In the case of punicalagin isomers, the content was significantly higher in conventional fresh juice. With these results, it can be concluded that organic pomegranate juices is a good option for pomegranate processing because of two main reasons: (1) the quality, functional and sensory, of fresh and processed juices is optimal, and (2) organic agricultural practices are considered to be more respectful with the environment. Since all fruits used in this study are from the same cultivar, *Mollar de Elche*, the main differences were attributed to the farming type, showing that phenolic compounds were not affected by conventional or organic farming, whereas pomegranate juice antioxidant capacity decreased by using chemical fertilizers and pesticides.

AUTHOR CONTRIBUTIONS

N.N.J., M.C.L., Á.A.C.B., F.H., and Á.C.S. planned and designed the experiments; N.N.J. and M.C.L. performed the experiments; N.N.J., M.C.L. and Á.C.S. analyzed the data; N.N.J., M.C.L., Á.A.C.B., F.H., and Á.C.S. wrote the manuscript and Á.C.S. edited the manuscript.

CONFLICTS OF INTEREST

The authors declare no conflict of interest.

REFERENCES

1. Raigón, M.D.; Dominguez-Guento, A.; Tortosa, A.; Carot-Sierra, J.M. Comparación de Rendimiento en Zumo y Contenido en Vitamina C de Diversas Variedades de Cítricos, Cultivadas Bajo Sistemas Ecológicos y Convencionales. Available online: http://fci.uib.es/Servicios/libros/conferencias/seae/Comparacion-de-rendimiento-en-zumo-y-contenido-en.cid221919 (accessed on 14 January 2015). (In Spanish).

2. Pacini, C.; Wossink, A.; Giense, G.; Vazzana, C.; Huirne, R. Evaluation of sustainability of organic, integrated and conventional farming systems: A farm and field-scale analysis. *Agric. Ecosyst. Environ.* **2003**, *95*, 273–288.

3. Madaula, F. Alimentos y salud: una relación de calidad. *Savia* **1996**, *3*, 36–38. (In Spanish).

4. Tarozzi, A.S.; Hrelia, C.; Angeloni, F.; Morroni, P.; Biagi, M.; Guardigli, G.; Cantell-Forti, P.; Hrelia, P. Antioxidant effectiveness of organically and non-organically grown red oranges in cell culture systems. *Eur. J. Nutr.* **2006**, *45*, 152–158.

5. Macheix, J.J.; Fleuriet, A.; Billiot, J. Changes and metabolism of phenolic compounds in fruits. In*Fruit Phenolics*; Macheix, J.J., Fleuriet, A., Billiot, J., Eds.; CRC Press: Boca Raton, FL, USA, 1990; pp. 149–221.

6. Melgarejo, P.; Salazar, D.M. *Tratado de Fruticultura para Zonas Áridas y Semiáridas*; Mundi-Prensa: Madrid, Spain, 2003; Volume II, p. 430. (In Spanish)

7. Nuncio-Jáuregui, N.; Calín-Sánchez, Á.; Vázquez-Araujo, L.; Pérez-López, A.J.; Frutos-Fernández, M.J.; Carbonell-Barrachina, Á.A. Processing pomegranates for juice and

impact on bioactive components. In *Processing and Impact on Active Components in Food*; Elsevier: London, UK, 2015; Chapter 76; pp. 629–636.

8. Gil, M.I.; Tomás-Barberán, A.; Hess-Pierce, B.; Holcroft, D.M.; Kader, A.A. Antioxidant activity of pomegranate juice and its relationship with phenolic composition and processing. *J. Agric. Food Chem.* **2000**, *48*, 4581–4589.

9. Seeram, N.P.; Aviram, M.; Zhang, Y.; Henning, S.M.; Feng, L.; Dreher, M.; Heber, D. Comparison of antioxidant potency of commonly consumed polyphenol-rich beverages in the United States. *J. Agr. Food Chem.* **2008**, *56*, 1415–1422.

10. Basu, A.; Penugonda, K. Pomegranate juice: A heart-healthy fruit juice. *Nutr. Rev.* **2009**, *67*, 49–56.

11. Malik, A.; Afaq, F.; Sarfaraz, S.; Adhami, V.; Syed, D.; Mukhtar, H. Pomegranate fruit juice for chemoprevention and chemotherapy of prostate cancer. *Proc. Natl. Acad. Sci. USA* **2005**, *102*, 14813–14818.

12. Sánchez, M.; Grande, I.; Gil, J.M.; Gracia, A. Evaluación el Potencial de Mercado de los Productos de Agricultura Ecológica. Available online: http://www.unavarra.es/personal/m_sanchez/evaluac.pdf (accessed on 13 January 2015). (In Spanish).

13. Lairon, D. Nutritional quality and safety of organic food. A review. *Agron. Sustain. Dev.* **2010**, *30*, 33–41.

14. Rembialkowska, E. Quality of plant products from organic agriculture. *J. Sci. Food Agric.* **2007**, *87*, 2757–2762.

15. Dangour, A.D.; Dodhia, S.K.; Hayter, A.; Allen, E.; Lock, K.; Uauy, R. Nutritional quality of organic foods: A systematic review. *Am. J. Clin. Nutr.* **2009**, *90*, 680–685.

16. Bourn, D.; Prescott, J. A comparison of the nutritional value, sensory qualities and food safety of organically and conventionally produced foods. *Crit. Rev. Food Sci. Nutr.* **2002**, *42*, 1–34.

17. Chong, C.H.; Law, C.L.; Figiel, A.; Wojdyło, A.; Oziembłowski, M. Colour, phenolic content and antioxidant capacity of some fruits dehydrated by a combination of different methods. *Food Chem.***2013**, *141*, 3889–3896.

18. Gao, X.; Ohlander, M.; Jeppsson, N.; Bjork, L.; Trajkovski, V. Changes in antioxidant effects and their relationship to phytonutrients in fruits of sea buckthorn (*Hippophae rhamnoides* L.) during maturation. *J. Agric. Food Chem.* **2000**, *48*, 1485–1490.

19. Brand-Williams, W.; Cuvelier, M.E.; Berset, C. Use of free radical method to evaluate antioxidant activity. *Lebensm. Wiss. Technol.* **1995**, *28*, 25–30.

20. Calín-Sánchez, Á.; Figiel, A.; Hernández, F.; Melgarejo, P.; Lech, K.; Carbonell-Barrachina, Á.A. Chemical composition, antioxidant capacity, and sensory quality of pomegranate (*Punica granatum* L.) arils and rind as affected by drying method. *Food Bioprocess. Tech.* **2013**, *6*, 1644–1654.

21. Rice-Evans, C.A.; Miller, N.J.; Paganga, G. Antioxidant properties of phenolic compounds. *Trends Plant Sci.* **1997**, *2*, 152–159.

22. Mena, P.; García-Viguera, C.; Navarro-Rico, J.; Moreno, D.A.; Bartual, J.; Saura, D.; Martí, N. Phytochemical characterisation for industrial use of pomegranate (*Punica granatum* L.) cultivars grown in Spain. *J. Sci. Food Agric.* **2011**, *91*, 1893–1906.

23. Roussos, P.A. Phytochemicals and antioxidant capacity of orange (*Citrus sinensis* (l.) Osbeck cv. Salustiana) juice

produced under organic and integrated farming system in Greece. *Sci. Hort.* **2011**, *129*, 253–258.

24. Dani, C.; Oliboni, L.; Vanderlinde, R.; Bonatto, D.; Salvador, M.; Henriques, J. Phenolic content and antioxidant activities of white and purple juice manufactured with organically- or conventionally produced grapes. *Food Chem. Toxicol.* **2007**, *45*, 2574–2580.

25. Yildiz, H.; Bozkurt, H.; Icier, F. Ohmic and conventional heating of pomegranate juice: Effects on rheology, color, and total phenolics. *Food Sci. Technol. Intl.* **2009**, *15*, 503–512.

26. Hernández, F.; Melgarejo, P.; Tomás-Barberán, F.A.; Artés, F. Evolution of juice anthocyanins during ripening of new selected pomegranate (*Punica granatum*) clones. *Eur. Food Res Technol.* **1999**, *210*, 39–42.

27. Gastol, M.; Domagala-Swiatkiewicz, I.; Krosniak, M. Organic *verses* conventional—A comparative study on quality and nutritional value of fruit and vegetable juices. *Biol. Agric. Hortic.* **2011**, *27*, 310–319.

28. Asami, D.K.; Hong, Y.J.; Barrett, D.M.; Mitchell, A.E. Comparison of the total phenolic and ascorbic acid content of freeze-dried and air-dried marionberry, strawberry and corn grown using conventional, organic and sustainable agricultural practices. *J. Agric. Food Chem.* **2003**, *51*, 1237–1241.

29. Lu, J.; Ding, K.; Yuan, Q. Determination of punicalagin isomers in pomegranate husk. *Chromatographia* **2008**, *68*, 303–306.

30. Vegara, S.; Mena, P.; Martí, N.; Saura, D.; Valero, D. Approaches to understanding the contribution of

anthocyanins to the antioxidant capacity of pasteurized pomegranate juice. *Food Chem.* **2013**, *141*, 1630–1636.

31. Carvajal de Pabón, L.M.; Yahia, E.H.; Cartagena, R.; Peláez, C.; Gaviria, C.A.; Rojano, B.A. Antioxidant capacity of two Fragaria x ananassa (Weston) Duchesne (strawberry) varieties subjected to variations in vegetal nutrition. *Revista Cubana de Plantas Medicinales* **2012**, *17*, 37–53.

32. Carbonell-Barrachina, Á.A.; Calín-Sánchez, Á.; Bagatar, B.; Hernández, Fca.; Legua, P.; Martínez-Font, R.; *et al.* Potential of Spanish sour-sweet pomegranates (cultivar C25) for the juice industry.*Food Sci. Technol. Int.* **2012**, *18*, 129–138.

33. Onsekizoglu, P. Production of high quality clarified pomegranate juice concentrate by membrane processes. *J. Membr. Sci.* **2013**, *442*, 264–271.

7

CHAPTER

Pomegranate juice enhances healthy lifespan in Drosophila melanogaster: an exploratory study

Subramani Paranthaman Balasubramani[1,2], Jayaram Mohan[3], Arunita Chatterjee[3], Esha Patnaik[3], Subrahmanya Kumar Kukkupuni[1,2], Upendra Nongthomba[3]* and Padmavathy Venkatasubramanian[1]*

[1]*Foundation for Revitalisation of Local Health Traditions (FRLHT), Bangalore, India*

[2]*Manipal University, Manipal, India*

[3]*Department of Molecular Reproduction, Development and Genetics, Indian Institute of Science, Bangalore, India*

ABSTRACT

Exploring innovative ways to ensure healthy aging of populations is a pre-requisite to contain rising healthcare costs. Scientific research into the principles and practices of traditional medicines can provide new insights and simple solutions to lead a healthy life. Rasayana is a dedicated branch of Ayurveda (an Indian medicine) that deals with methods to increase vitality and delay aging through the use of diet, herbal supplements, and other lifestyle practices. The life-span and health-span enhancing actions of the fruits of pomegranate (*Punica granatum* L.), a well-known Rasayana, were tested on *Drosophila melanogaster* (fruitfly) model. Supplementation of standard corn meal with 10% (v/v) pomegranate juice (PJ) extended the life-span of male and female flies by 18 and 8%, respectively. When male and female flies were mixed and reared together, there was 19% increase in the longevity of PJ fed flies, as assessed by MSD, the median survival day (24.8). MSD for control and resveratrol (RV) groups was at 20.8 and 23.1 days, respectively. A two-fold enhancement in fecundity, improved resistance to oxidative stress (H_2O_2 and paraquat induced) and to *Candida albicans* infection were observed in PJ fed flies. Further, the flies in the PJ fed group were physically active over an extended period of time, as assessed by the climbing assay. PJ thus outperformed both control and RV groups in the life-span and health-span parameters tested. This study provides the scope to explore the potential of PJ as a nutraceutical to improve health span and lifespan in human beings.

Keywords: pomegranate, anti-aging, rasayana, ayurveda, *Drosophila*

INTRODUCTION

Global statistics predict 223% increase in the number of elderly (aged >60 years) accounting for about 1.2 billion between 1970 and 2025 (1). Ways to ensure healthy aging while increasing life-span are heatedly debated and researched in public health circles (2). This is because long-term institutional care of the elderly for chronic age-related conditions such as arthritis and dementia has led to steep increases in healthcare costs, yet reduced the quality of life for the patients (3, 4). Countries have started recognizing that promoting self-reliant wellness strategies among communities would not only save the exchequer from substantial financial burden due to illness care but also have a population that is healthy and productive (5). Increasing evidence through longitudinal studies show that simply adopting appropriate diet and lifestyle can even reverse chronic conditions like coronary atherosclerosis (6).

Complementary and alternative medicines like ayurveda, yoga, and traditional Chinese medicine prescribe several interventions for prevention of diseases and promotion of health that can be scientifically explored for wellness. *Rasayana* is a special branch of Ayurveda, which deals with methods of rejuvenation such as dietary recipes and regimen, herbal and mineral supplements, and health-promoting lifestyle that are said to enhance quality of life and delay aging (7). Traditional literature describes Rasayanas as methods of reversing naturally occurring senility and improving mental competence (8), increasing immunity against diseases and providing vitality and luster to the body (7). As per Ayurveda, Rasayanas act on the body by (i) improving the process of digestion and metabolism (*Agni*), leading to better bio-availability, (ii) increasing the micro-circulation of nutrients through the body channels (*Srotas*), and (iii) enhancing the nutritive value of the plasma that gets generated (*Rasadhatu*) (7, 9). Several studies have indicated Rasayanas to be anti-oxidants and immunomodulators (10).

Ayurveda identifies several fruits and herbs including the Indian Gooseberry (*Phyllanthus emblica*), Pomegranate (*Punica granatum*), Drumsticks (*Moringa oleifera*), Long pepper (*Piper longum*), etc., as Rasayanas (7, 8, 10). Understanding the scientific basis behind the concept of Rasayana and validation of the life promoting ability of commonly available Rasayana fruits and herbs can help establish holistic and simple ways to wellness through diet.

The potential therapeutic properties of pomegranate, a commonly available fruit, have been studied for wide-ranging applications including cardiovascular disease, diabetes, dental conditions, erectile dysfunction, protection from ultraviolet (UV) radiation, infant brain ischemia, prevention of cancer, Alzheimer's disease, male infertility, arthritis, and obesity (11–13). Pomegranate is considered as a symbol of life, health, longevity, fecundity, intellect, immortality, and spirituality by Unani and Traditional Chinese Medical systems as well (14). Ayurvedic texts describe pomegranate as a "wholesome" (*Pathya*) fruit with several properties including tissue generation and development and strength promotion (15). It is particularly indicated for use in cardiac and anemic conditions (16).

Aging is a complex physiological phenomenon, particularly because it manifests itself over a wide range of biological systems, tissues, and functions. Aging pathways are highly conserved across species including *Saccharomyces cerevisiae*, *Caenorhabditis elegans*, *D. melanogaster*, and rodents (17, 18). *Drosophila* (commonly known as fruit fly) has many advantages for studying the biology of aging. For example, flies show physiological, genetic, and anatomical similarities with human (17). Owing to short-life-span and ease of handling, the fruit flies are particularly ideal for conducting prevention and wellness studies, which would otherwise take several years in human beings (17). Fruits like cranberry (19), blueberry (20), orange, lemon (21), apple polyphenols (22), and cocoa (23) have been shown to enhance lifespan in *Drosophila* model. A few Ayurvedic formulations have also been tested for their bioactive potential in the *Drosophila* model (24, 25).

Pomegranate, a member of the Punicaceae family is native to the Himalayas. However, it is also under commercial cultivation in other parts of India and the world (12). It contains polyphenols such as ellagitannins and anthocyanins, as well as phenolic acids, fatty acids, alkaloids, and a variety of volatile compounds (26). Ellagitannins are among the most prevalent compounds present in pomegranate and may be responsible for various bioactivities (27). The juice contains ellagic acid, gallic acid, anthocynidins, flavan-3-ols, straight chain fatty acids, citric acid, and malic acid besides glucose, fructose, and sucrose as the major sugars (26). Pomegranate has been shown to have anti-oxidant, anti-inflammatory, lipid regulation, gastroprotective, anti-diarrhea, anti-cancer, anti-viral, anti-angiogenesis, anti-diabetic, and anti-atherosclerotic activities (26, 27).

The Rasayana actions of pomegranate, namely, enhancement of life-span, reproductive ability, stress tolerance, resistance to infection, and stamina were compared between control, resveratrol (RV), and pomegranate juice (PJ) supplemented *D. melanogaster*. Our hypothesis was that pomegranate supplementation would significantly improve the performance of the flies on all counts, as compared to the control.

MATERIALS AND METHODS

Drosophila Strain, Media, and Culture Conditions

Drosophila melanogastor wild type, *Canton-S* strain, was used for this study. The basal media for culturing the flies consisted of corn flour (60 g), sugar (20 g), D-glucose (20 g), dry yeast (15 g), and agar (8 g) per liter. After cooking, media was allowed to cool to room temperature and then propionic acid (4 ml), benzoic acid (0.7 g in 5 ml of ethanol), orthophosphoric acid (0.6 ml) were added and mixed thoroughly. Media cooked as slurry was poured into the bottles or vials and was allowed to solidify before transfer of adult flies or pupae. Stock flies were maintained in bottles while culture vials were used for maintenance of smaller populations or experimental groups. Flies were allowed to feed *ad libitum* and grown at 25°C with 60% humidity in 12 h day–night cycle.

Preparation of Pomegranate Juice and Resveratrol Stock

Pomegranate juice was prepared by hand crushing the arils of fresh fruits through sterile muslin cloth. The juice was passed through a filter (0.2 μm; Millex, Millipore, Germany) and stored as aliquots of different volume (5, 15, and 50 ml) in screw capped tubes at −80°C until use. Freezing did not significantly alter the high-performance liquid chromatography (HPLC) profile (data not shown). RV (Sigma-Aldrich, St. Louis, MO, USA) stock (100 mM) was prepared by dissolving 22.8 mg in 1 ml of ethanol. Both PJ and RV were added to the cooked and warm basal media before dispensing into vials.

Lifespan Extension Experiment

Age matched (2–3 days) flies were generated by removing all adult flies in the bottles. Once a synchronous culture was generated, sex separated flies were transferred into vials containing basal medium supplemented with 0.1, 1, 5, 10, and 15% (v/v) of PJ. Flies were transferred to fresh media every third day. Dead flies were counted and removed daily until the death of the last fly in each vial. PJ at 15% v/v could not be used since at that concentration the medium would not set.

Longevity and Fecundity Analysis

Synchronous cultures of 2–3-day-old files were obtained as described earlier and transferred into vials containing basal medium supplemented with 10% of PJ (v/v) or 200 μM RV and control. Each group including the control had 10 vials each with 14–20 flies per vial with equal sex ratio. Flies were transferred to fresh media every third day. Dead flies were counted and removed daily throughout the experiment. New flies emerging in the emptied vials were counted daily for 10 days from the day of the emergence of the first fly in each experimental group.

Gustatory Assay

The method described by Lee et al. (28) was adopted for gustatory assay. Adult flies were reared for 20 or 40 days in media supplemented with 10% PJ or 200 μM RV. A basal media control was also maintained. To perform a feeding assay, after starving the flies for 2 h, 30 male or female flies from each experimental group were transferred into the vials containing the specific diets with bromophenol blue dye (0.05% wt/vol) (Sigma-Aldrich, St. Louis, MO, USA). After 10 min of feeding, the fed flies were etherized, washed with phosphate-buffered saline (PBS), and homogenized in 1 ml of distilled water. The absorbance of 100 times diluted supernatant was measured at 595 nm using a spectrophotometer (Bio-Rad, CA, USA).

Resistance Against Hydrogen Peroxide and Paraquat Induced Stress

The method described by Peng et al. (22) was used for studying the fly's ability to resist free radical stress induced by H_2O_2(Fisher Scientific, Mumbai, India) and paraquat (Sigma-Aldrich, St. Louis, MO, USA). H_2O_2 generates hydroxyl radical (OH*) while paraquat produces superoxide (O*) radicals. Twenty days old male and female flies from control, RV (200 μM), and PJ (10%) fed were starved for 2 h and transferred to separate vials (n = 50; 10 flies/vial) containing 5% H_2O_2 or 20 mM paraquat prepared with 5% sucrose on saturated tissue paper mat. Dead flies were counted every 4 h until the death of the last fly.

Resistance Against Infection

The method described by Apidianakis and Rahme (29) was followed with slight modifications. Antigen was prepared by suspending an overnight culture of *C. albicans* SC5314 in sterile distilled water (2×10^8 cells/ml – as counted using a hemocytometer; data not shown). Flies (20 days old) were anesthetized with CO_2. Infection was induced with a sterile tungsten needle (0.01 mm diameter) dipped in the *C. albicans* suspension by gently pricking in the thoracic region. Flies were returned to their respective food vials and were incubated at 25°C with 12 h day–night cycle until 50% of flies died in any of the groups. Control flies were pricked with needle dipped in sterile distilled water.

Climbing Assay

On day 20 and 40, locomotor function of fruit flies was assessed using the climbing or negative geo-taxis assay as reported by Bahadorani and Halliker (23) with slight modifications. In brief, 20 flies/trial were placed in the bottom of a measuring cylinder and given 20 s to climb up. At the end of each trial, the number of flies that climbed up to a vertical distance of >8 cm was recorded. Each trial was repeated three times.

Statistical Analysis

The day when 50% of the total flies in each of the experimental group survived was calculated as the median survival day. Log rank test[1] was used for survival analysis while Wilcoxon rank sum test[2] was used for comparing the life-span extension (30). Student's *t*-test was used for comparison of means. ANOVA was employed while comparing more than two groups. *p*-Value <0.05 was considered as significant.

RESULTS

PJ Supplementation Extends Lifespan in *Drosophila*

To determine the lifespan extending potential in *Drosophila*, we supplemented the basal media with 0.1, 1, 5, 10, and 15% of PJ. Substitution of the media with 10% PJ significantly enhanced the median lifespan by 18.51 and 8% in male and female flies, respectively (Figure 1; Table 1). Last fly survival day was also enhanced by 2 days in the 10% PJ group. While the lower concentration tested (0.1, 1, and 5%) did not produce any significant changes in the life-span, 15% PJ substitution created a practical difficulty in solidification of the medium; therefore, this concentration was not considered. Thus, 10% PJ was used in all further experiments.

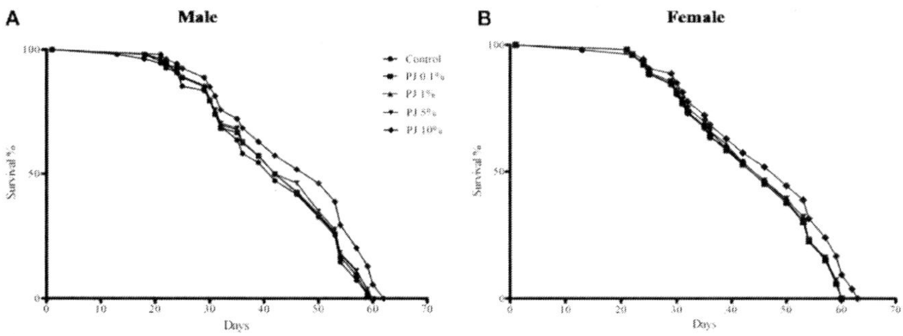

Figure 1. Life-span extension by different concentrations of pomegranate juice (PJ) in *Drosophila* (A) male flies and (B) female flies. Ten percent PJ substitution to the media significantly increased ($p < 0.00001$) the survival of the flies and extended their life-span by 2 days. In both the sexes 0.1, 1, or 5% PJ did not show any enhancement in the survival.

PJ Supplementation Controls the "Trade-Off" between Fecundity and Longevity

As 10% PJ substitution to the media significantly enhanced life-span for both male and female flies separately, we investigated the effect of this concentration in the mixed group of male and female. RV at 200 μM dose was used as a positive control, based on the reports of Bass et al. (31).

Pomegranate juice enhanced the median survival of a mixed population of flies from 20.8 (as seen in control) to 24.8 days (PJ), i.e., PJ enhanced median life-span by 19.23% more than the control ($p < 0.00001$) (Figure 2A; Table 1). RV group on the other hand enhanced the median survival to 23.1 days, which was only 11% over the control group (Figure 2A; Table 1). The maximum life-span observed in the control and PJ groups was 61 days while that in the RV group, the last fly survived till 62nd day. Reproductive potential of an organism also indicates its physiological fitness. So, we monitored the number of flies emerging from each vial of PJ, RV, and control group. The data shown in Figure 2B indicate that PJ group produced significantly ($p < 0.0001$) more number of flies throughout the reproductive phase. Also, the new fly emergence data indicated that PJ extended the reproductive viability phase (Figure 2B).

Table 1. Statistical analysis of the survival curves using Log rank tests and Wilcoxon rank sum tests.

Experimental group	Number of flies tested (n)	Median survival day/hour	% Change	Chi square value @ 1 degree of freedom	Log rank (survival)	Wilcoxon (maximum life-span)
Lifespan extension						
Male						
Control	110	40.5				
PJ 0.1	108	42.0	3.7	0.7	$p = 0.396$	$p \leq 0.6421$
PJ 1	112	42.0	3.7	1.6	$p = 0.207$	$p \leq 0.5322$
PJ 5	108	42.0	3.7	3.9	$p = 0.047$	$p \leq 0.6208$
PJ 10	108	48.0	18.51	46.2	$p < 0.00001$*	$p \leq 0.2545$
Female						
Control	104	43.5				
PJ 0.1	106	43	−1.14	0.1	$p = 0.699$	$p \leq 0.6861$
PJ 1	112	43.5	0	0.1	$p = 0.720$	$p \leq 0.5084$
PJ 5	112	44	1.14	0.1	$p = 0.758$	$p \leq 0.5392$
PJ 10	108	47	8.0	19	$p < 0.00001$*	$p \leq 0.3439$
Longevity and fecundity						
Control	247	20.8				
RV	167	23.1	11.05	16.2	$p < 0.00001$*	$p \leq 0.6016$
PJ	164	24.8	19.23	48.7	$p < 0.00001$*	$p \leq 0.3506$
H_2O_2 stress survival						
Male						
Control	50	18.0				
RV	50	24.5	36.1	29.3	$p < 0.00001$*	$p \leq 0.4705$
PJ	50	28.0	55.5	1.6	$p < 0.00001$*	$p \leq 0.2602$
Female						
Control	50	19.0				
RV	50	26.5	39.4	22.5	$p < 0.00001$*	$p \leq 0.4237$
PJ	50	29.5	55.2	49.4	$p < 0.00001$*	$p \leq 0.1824$
Paraquat stress survival						
Male						
Control	50	15.5				
RV	50	17.0	9.6	7.4	$p = 0.0066$*	$p \leq 0.6295$
PJ	50	24.0	54.8	19.2	$p < 0.00001$*	$p \leq 0.4907$
Female						
Control	50	18.5				
RV	50	24.5	32.4	8.7	$p = 0.0032$*	$p \leq 0.4553$
PJ	50	27.0	45.9	30.3	$p < 0.00001$*	$p \leq 0.2856$

	n	Median survival time (MST) in hours	% MST increase	% of surviving flies	ANOVA	t-test
Resistance against infection						
Male						
Control	50	4.88		41	$p = 0.0013$*	
RV	50	>6	>23	55		$p \leq 0.05$*
PJ	50	>6	>23	64		$p \leq 0.05$*
Female						
Control	50	3.82		38	$p = 0.0012$*	
RV	50	>5	>31	59		$p \leq 0.05$*
PJ	50	>5	>31	78		$p \leq 0.05$*

*Significant

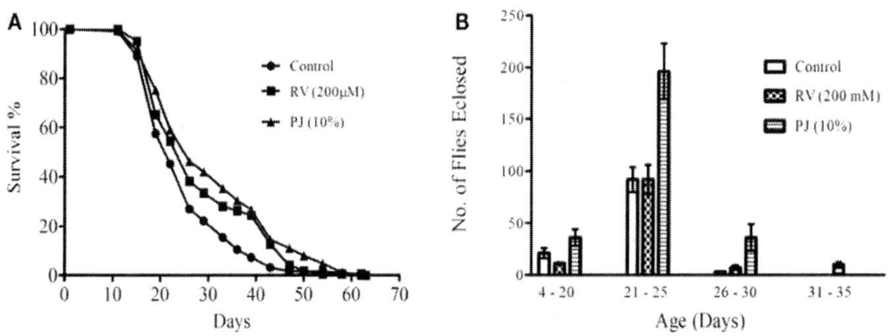

Figure 2. (A). Effect of pomegranate juice supplementation on *Drosophila* longevity. Male and female mixed culture flies were fed with 10% PJ (n = 164) or 200 μM RV (n = 167). A basal medium control (n = 247) was also maintained. Log rank test indicated significant ($p < 0.00001$) increase in the fly survival by both PJ and RV (Table 1). But PJ increased the median survival by 19.23% while it was 11.05% in the RV group. **(B)** Effect of 10% pomegranate juice supplementation on *Drosophila* fecundity at different age. The bars represent the total number of flies enclosed from each group at different age with SD. PJ group showed significant increase in the number of off-springs produced (*$p < 0.0001$). The graph also indicates an extended reproductively viable phase in the PJ fed flies. No significant difference in fecundity was observed with RV and control groups.

Drosophila has the Same Feeding Preference to PJ and RV Supplementation

Feeding behavior and nutritional constituents of a culture medium are important factors in the life-span determination of *D. melanogaster* (30). Bass et al. (31) has reported that RV enhances lifespan by dietary restriction in *Drosophila* and *C. elegans*. To ensure that the observed changes in the lifespan were due to PJ or RV supplementation and not because of starvation or dietary restriction, gustatory assay was performed, quantifying the food intake on 20- and 40-day-old flies using bromophenol blue dye. While there was no significant difference observed between PJ and RV fed groups, the feed intake was lower in both these groups as compared to that in the control ($p < 0.05$), particularly in the male flies (Figure 3A). This difference was not significant in the female flies. The feed intake of 40-day-old female flies was similar in control, PJ, and RV groups (Figure 3B). Also, there was a significant difference ($p < 0.05$) in food intake between the male and female flies in both PJ and RV groups (Figure 3).

PJ Enhances Tolerance to Free Radical Induced Stress

Ayurveda claims that Rasayanas improve the stress resistance capacity of organisms. Both H_2O_2 and paraquat produce free radicals and are lethal to flies. A significant ($p < 0.00001$) increase in the post H_2O_2 and paraquat exposure survival time was observed in both PJ and RV supplemented flies when compared to the control (Figures 4 and 5; Table 1). Analysis of median survival indicated an average 50% increase over the control group in both male and female flies in PJ. Though, RV was also found to protect against the free radical induced stress, its level of protection showed only about 30–35% increase in survival (Figures 4 and 5).

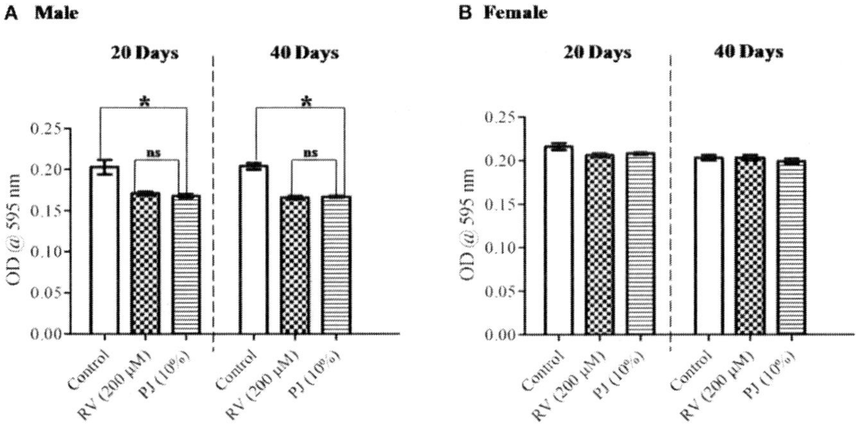

Figure 3. Gustatory assay with PJ (10%) or RV (200 μM) supplementation in 20 and 40 days old *Drosophila* flies.(A) Male flies in the PJ and RV group were found to have reduced feed intake when compared to control ($*p < 0.05$). However, there was no difference in food intake by PJ and RV groups. **(B)** Female flies did not show any significant difference in the food intake on both the days tested (ns, not significant).

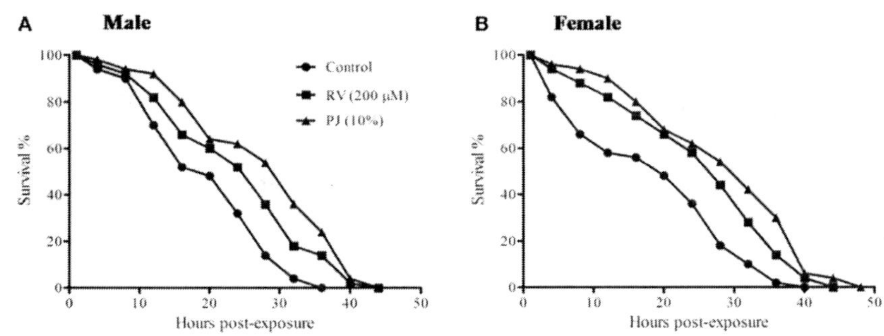

Figure 4. Survival against hydroxyl radical (OH*) produced by hydrogen peroxide exposure in 20 days old*Drosophila*. Both male **(A)** and female **(B)** flies fed with PJ (10%) showed approximately 50% increase in median survival ($p < 0.00001$) (Table 1). RV (200 μM) fed flies showed only about 30–35% increase in median survival.

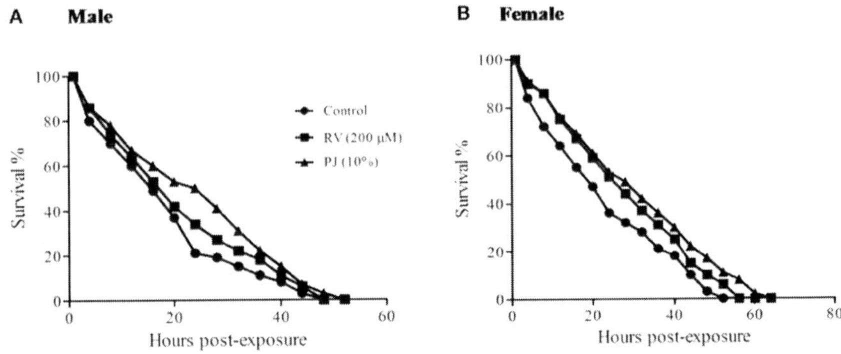

Figure 5. Survival against superoxide radical (O*) produced by paraquat exposure in 20 days old *Drosophila*. (A)PJ (10%) group showed 54% increase in median survival ($p < 0.00001$) while RV (200 µM) could increase only by 9.6% (Table 1).**(B)** Female flies fed with PJ showed 45.9% increase in median survival ($p < 0.00001$) while RV was able to increase the median survival by 32.4%.

PJ Protects *Drosophila* from *Candida albicans* Infection

Drosophila melanogaster has been used as a model to study host–fungal interaction and pathogenicity (32). Glittenberg et al. (33) established *Drosophila* as an alternative model for investigating the pathogenicity of *C. albicans*. Literature indicates that Rasayanas have immunomodulatory (*Vyadhikshamatwa*) potential (10). As proof of principle, we evaluated if PJ substitution offered protection to the flies from death because of *C. albicans* infection. The survival curve indicated a significant ($p \leq 0.05$) protective effect of PJ until 144 h post-infection. Even though, RV too protected the flies (>50%) from *C. albicans* infection, the magnitude of protection by PJ (>70%) was significantly higher. The control flies had only 40% survival (Figure 6). While the magnitude of protection differs with male and female flies, the protection trend was similar.

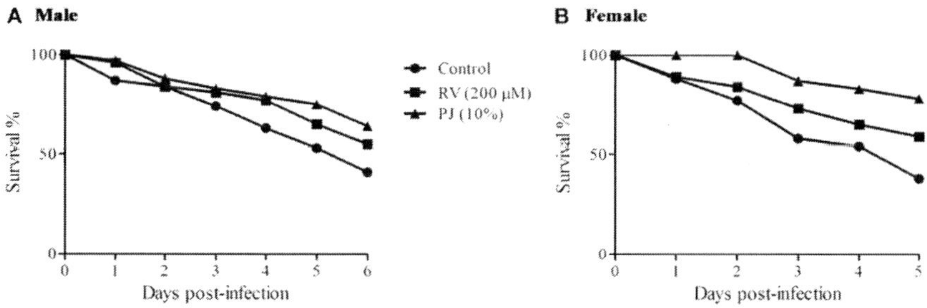

Figure 6. Pomegranate protects *Drosophila* **from** *Candida albicans* **infection**. Twenty days old male **(A)** and female **(B)**flies fed with PJ, RV, or control diet were infected with *Candida albicans*. In both the sexes, flies fed with PJ showed a significantly ($p \leq 0.05$) better survival response than the control and RV fed.

PJ Promotes Sustained Physical Performance in *D. melanogaster*

The success of healthy living lies in retention of normal physical activity. Rasayanas like pomegranate are said to improve energy, strength, and stamina (*Balya*) of organisms. The physical performance of the flies was measured using the climbing or negative geo-taxis assay. This assay was performed with 20 and 40 days old male and female flies fed on media supplemented with PJ or RV. Even though, significant difference among the groups was not observed in the percentage of flies climbing on 20th day, the PJ group retained the physical activity even on day 40 (aged flies), while the other groups showed severe decline ($p < 0.05$) (Figure 7). Thus, PJ was found to sustain good physical activity in terms of negative geo-taxis climbing activity during aging.

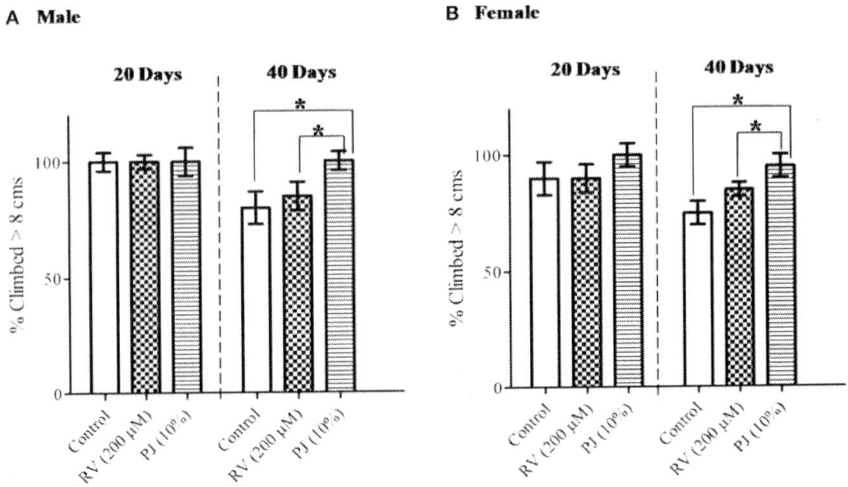

Figure 7. Sustained climbing performance of *Drosophila* flies fed with PJ. (A) Twenty days old male flies did not show any difference in their climbing performance irrespective of feed, but 40 days old PJ fed flies were able to retain their physical performance ($p < 0.0001$) then the RV and control group. **(B)** Female flies at both 20 and 40 days showed significantly higher ($p < 0.0001$) climbing performance then the RV and control group. PJ group was able to resist the age-related physical activity decline observed in the RV and control groups.

DISCUSSION

Developing scientifically validated, culturally acceptable, appropriate guidelines for healthy eating is recognized as one of the thrust areas for reducing age-related functional decline (1). An Australian longitudinal analysis to guide health promotion concludes that health, life style, and gender influences healthy aging (34). In Ayurveda, Rasayanas are said to improve immunity, impart vitality, are good aphrodisiacs and are considered to delay the aging process (8, 10). Pomegranate has been mentioned in Ayurveda as a Rasayana with the above properties, which was the reason for selecting the fruit for our study. We have shown that PJ fed flies live and reproduce longer (Figure 2B), are better protected against stress and infection (Figures4–6) and are active even when aged (Figure 7). According to the antagonistic pleiotrophy theory, higher levels of reproduction are negatively correlated with survival. This concept of "trade-off" between longevity and reproduction or "cost of reproduction" has been widely accepted and demonstrated in a number of experimental studies (17). However, our observation in the current experimental setup indicated that PJ is capable of simultaneously enhancing fertility as well as survival. Reduction in the fertility of individuals has been considered as one of the reasons for accelerated graying in population (1). Ayurvedic texts indicate Rasayanas as holistic interventions that reduce age-related senility while maintaining physical and biological performance of the organism (7). This study opens up the scope of scientifically exploring Rasayanas to identify methods and to counter the burning contemporary issue of healthy aging.

Studies have shown that elevated resistance to various environmental stresses, including oxidative stress is one of the phenotypes of long lived drosophila mutants (22). When the protecting ability of PJ and RV was assessed against H_2O_2 and paraquat induced oxidative stress, we observed that PJ could play a role in ameliorating the free radical induced damage, which is one of the reasons for aging. The anti-oxidant potential of pomegranate demonstrated by Faria and Calhau (12) support our findings. Cranberry extract, apple polyphenols, curcumin, are some of the fruits and herbs that have been reported to impart protection against H_2O_2 and paraquat induced free radicals in *Drosophila* (19, 22, 28). Stress not only affects older individuals but it is also predisposed to cause premature aging (35). Free radicals and oxidative stress are increasingly being accepted as a major player in several disease processes and accelerated aging (36). Flavonoids and polyphenols in fruits like cranberry and blueberry have been shown to protect cells from oxidative stress induced damage caused by pro-inflammatory mediators (37). The polyphenol content in pomegranate may be responsible for imparting resistance against free radical induced stress (26). While fruit powders of blueberry and cranberry impart health benefits to flies at 2–3% w/v, orange juice was found to be effective at higher concentrations (3–50% v/v) (19, 20). Purified apple polyphenols (10 mg/ml of fly diet) were found to significantly enhance the health and life-span of flies. The concentration of polyphenols was estimated to be 0.4/100 kg of apple pomace (22).

Drosophila, although devoid of an adaptive immune system, harbors an innate immune response with striking similarities to the mammalian defense mechanisms. The *Drosophila* innate immune response uses pattern recognition receptors to activate phagocytosis by plasmatocytes, proteolytic clotting cascades in the hemolymph and production of specific antimicrobial peptides (AMPs) in the fly fat body (32). The expression of specific anti-fungal peptides is controlled by a signaling pathway orchestrated by the toll-like receptors (TLR) similar to human beings (38). The exact mechanism by which PJ exerted protection to the flies is yet to be elucidated. Infectious diseases have been posing a major threat to the elderly population because of age-related decline in immune functions (39). Such conditions can be overcome by use of immune-stimulators. Several Rasayanas like pomegranate can be simple interventions to enhance immunity. *Drosophila* has been established as a model to also study age-related impairment in physical movement (40). Present study indicates that pomegranate feeding can sustain physical movement in aged flies. The extended reproductive phase (Figure 2B) and sustained physical activity (Figure 7) in the flies can be considered as direct indicators of healthy life-span.

The current study corroborates earlier reports that RV enhances lifespan in *Drosophila* (18, 31, 41). In this study, we have observed that RV increased median life-span of *Drosophila* by 11% (Figure 2A). RV substitution imparted comparatively lesser protective ability against H_2O_2, paraquat, and infection stress to the flies when compared to PJ fed group. RV did not significantly alter the fecundity of the flies as compared to the control. Estimation of the feed intake showed a slightly reduced feed intake quantity (Figure 3) in both PJ and RV supplemented groups when compared to the control. Earlier reports also indicate the possible dietary restriction by RV (31). In our experiments, this observation was more marked in the male flies than in the females. Ayurveda literature on pomegranate mentions that pomegranate provides satiation (*Trupti*), i.e., the organism is satisfied with lesser intake of food when supplemented with PJ (15, 16). While there was no significant difference in the feed intake between RV and PJ fed groups, the latter outperformed the former in all life-span and health-span parameters tested.

Free radical induced stress tolerance, improved infection survival, and climbing performance assays show that, pomegranate supplementation helps in reducing age-related functional decline in the flies. This study justifies the Ayurvedic claims that pomegranate is a Rasayana by being a multi-functional health promoter in fruit fly model. Given the exploratory nature of our study and the large number of comparisons made, the results are certainly very interesting; and it has opened up avenue for further research. Further research needs to be done to study the effectiveness of PJ supplementation in human beings at the community setting. Scientific exploration of other Rasayanas as quality-of-life enhancers is a potential area for research. Studying the mode of action of PJ, identification of bioactive components and molecular targets would be of value to understanding aging and anti-aging process.

Drosophila has been a strong candidate as a model in contemporary aging research. *D. melanogaster* has also been used as a model by researchers to study Rasayanas. An insect specific Rasayana was developed that was found to increase lifespan by 50% in *Drosophila* model (24). Dwivedi et al. (25) reported that Amalaki Rasayana, an herbal formulation and Rasa-Sindoor an organo-metallic derivative of mercury were capable of increasing life-span and fecundity in fruit flies.

Aging population and the associated physical, physiological, and neurological deficiencies are being recognized as the major challenges for healthcare in the twenty-first century. While people are living longer; they are aging prematurely (42). Scientific exploration of Rasayana can provide affordable ways to delay aging and maintain good health.

AUTHOR CONTRIBUTIONS

Padmavathy Venkatasubramanian conceptualized the transdisciplinary research methodology, obtained fund support, and edited the manuscript. Upendra Nongthomba designed *Drosophila* based experiments, analyzed results, and edited manuscript. Subramani Paranthaman Balasubramani performed the experiments, analyzed, interpreted results, prepared, and edited the manuscript. Jayaram Mohan maintained the *Drosophila* stocks and assisted Subramani Paranthaman Balasubramani in experimentation. Esha Patnaik and Chatterjee Arunita standardized the infection and stress experimentation protocol and analysis of results. Subrahmanya Kumar Kukkupuni provided the Ayurveda knowledge.

CONFLICT OF INTEREST STATEMENT

The authors declare that the research was conducted in the absence of any commercial or financial relationships that could be construed as a potential conflict of interest.

ACKNOWLEDGMENTS

Padmavathy Venkatasubramanian, Subramani Paranthaman Balasubramani, and Subrahmanya Kumar Kukkupuni wish to thank the Department of Science and Technology, Government of India for the financial support received under the Drugs and Pharmaceuticals Research Program (DPRP) scheme. Thanks to Dr. Venugopalan Nair for his Ayurvedic inputs and Prof. Nagarajan for his suggestions. Authors wish to thank Prof. Parag Sadhale (late) for providing *C. albicans* strain. Thanks to Prof. P. Kondaiah, MRDG and The Director, Indian Institute of Science for extending the facility. The facility and encouragement provided by Shri. Darshan Shankar, FRLHT are thankfully acknowledged.

FOOTNOTES

1. ^http://bioinf.wehi.edu.au/software/russell/logrank/

2. ^http://www.fon.hum.uva.nl/Service/Statistics/Wilcoxon_Test.html

REFERENCES

1. World Health Organisation. *Active Ageing: A Policy Framework*. Geneva: Non-communicable Disease Prevention and Health Promotion, Ageing and Life Course (2002). WHO/NMH/NPH/02.8. Available from: http://whqlibdoc.who.int/hq/2002/who_nmh_nph_02.8.pdf

2. Arai H, Ouchi Y, Yokode M, Ito H, Uematsu H, Eto F, et al. Towards the realization of a better aged society: messages from gerontology and geriatrics. *Geriatr Gerontol Int* (2012) **12**:16–22. doi: 10.1111/j.1447-0594.2011.00776.x

3. Chatterji S, Kowal P, Mathers C, Naidoo N, Verdes E, Smith JP, et al. The health of aging populations in China and India. *Health Aff* (2008) **27**:1052–63. doi:10.1377/hlthaff.27.4.1052

4. Avendano M, Glymour MM, Banks J, Mackenbach JP. Health disadvantage in US adults aged 50 to 74 years: a comparison of the health of rich and poor Americans with that of Europeans. *Am J Public Health* (2009) **99**:540–8. doi:10.2105/AJPH.2008.139469

5. National Health Services. *From Illness to Wellness – Achieving Efficiencies and Improving Outcomes*. London: NHS Confederation (2011). Available from:http://www.nhs-chaplaincy-spiritualcare.org.uk/NationalHealthService/illness_to_wellness_241011.pdf

6. Daubenmier JJ, Weidner G, Sumner MD, Mendell N, Merritt-Worden T, Studley J, et al. The contribution of changes in diet, exercise, and stress management to changes in coronary risk in women and men in the multisite cardiac lifestyle intervention program. *Ann Behav Med* (2007) **33**:57–68. doi:10.1207/s15324796abm3301_7

7. Udapa MH. *Comprehensive Kaayachikitsa and Principles of Ayurveda*. Bangalore: Laveena Publications (2004).

8. Sharma PV, editor. *Susruta-Samhita*. (Vol. **1**). Varanasi: Chaukhambha Visvabharati (2004).

9. Singh RH, Narsimhamurthy K, Singh G. Neuronutrient impact of ayurvedic rasayana therapy in brain aging. *Biogerontology* (2008) **9**:369–74. doi:10.1007/s10522-008-9185-z

10. Balasubramani SP, Venkatasubramanian P, Kukkupuni SK, Patwardhan B. Plant-based rasayana drugs from ayurveda. *Chin J Integr Med* (2011) **17**:88–94. doi:10.1007/s11655-011-0659-5

11. Jurenka JS. Therapeutic applications of pomegranate (*Punica granatum* L.): a review. *Altern Med Rev* (2008) **13**:128–44.

12. Faria A, Calhau C. The bioactivity of pomegranate: impact on health and disease. *Crit Rev Food Sci Nutr* (2011) **51**:626–34. doi:10.1080/10408391003748100

13. Johanningsmeier SD, Harris GK. Pomegranate as a functional food and nutraceutical source. *Annu Rev Food Sci Technol* (2011) **2**:181–201. doi:10.1146/annurev-food-030810-153709

14. Mahdihassan S. Outline of the beginnings of alchemy and its antecedents. *Am J Chin Med* (1984) **12**:32–42. doi:10.1142/S0192415X84000039

15. Chunekar KC, editor. *Bhavaprakasha Nighantu*. Varanasi: Chaukhambha Bharati Academy (2004).

16. Dash B. *Materia Medica of Ayurveda based on Madanapala's Nighantu*. New Delhi: B. Jain Publishers (P) Ltd (1991).

17. Iliadi KG, Knight D, Boulianne GL. Healthy aging – insights from *Drosophila*. *Front Physiol* (2012) **3**:106. doi:10.3389/fphys.2012.00106

18. Johnson SC, Rabinovitch PS, Kaeberlein M. mTOR is a key modulator of aging and age-related disease. *Nature* (2013) **493**:338–45. doi:10.1038/nature11861

19. Wang C, Yolitz J, Alberico T, Laslo M, Sun Y, Wheeler CT, et al. Cranberry interacts with dietary macronutrients to promote healthy aging in *Drosophila*. *J Gerontol A Biol Sci Med Sci* (2014) **69**:945–54. doi:10.1093/gerona/glt161

20. Peng C, Zuo Y, Kwan KM, Liang Y, Ma KY, Chan HY, et al. Blueberry extract prolongs lifespan of *Drosophila melanogaster*. *Exp Gerontol* (2012) **47**:170–8. doi:10.1016/j.exger.2011.12.001

21. Fernandez-Bedmar Z, Anter J, de La Cruz-Ares S, Munoz-Serrano A, Alonso-Moraga A, Perez-Guisado J. Role of citrus juices and distinctive components in the modulation of degenerative processes: genotoxocity, antigenotoxicity, cytotoxicity and longevity in *Drosophila*. *J Toxicol Environ Health A* (2011) **74**:1052–66. doi:10.1080/15287394.2011.582306

22. Peng C, Chan HY, Huang Y, Yu H, Chen ZY. Apple polyphenols extend the mean lifespan of *Drosophila melanogaster*. *J Agric Food Chem* (2011) **59**:2097–106. doi:10.1021/jf1046267

23. Bahadorani S, Hilliker AJ. Cocoa confers life span extension in *Drosophila melanogaster*. *Nutr Res* (2008) **28**:377–82. doi:10.1016/j.nutres.2008.03.018

24. Priyadarshini S, Ashadevi JS, Nagarjun V, Prasanna KS. Increase in *Drosophila melanogaster* longevity due to rasayana diet: preliminary results. *J Ayurveda Integr Med* (2010) **1**:114–9. doi:10.4103/0975-9476.65085

25. Dwivedi V, Anandan EM, Mony RS, Muraleedharan TS, Valiathan MS, Mutsuddi M, et al. In vivo effects of traditional ayurvedic formulations in *Drosophila melanogaster* model relate with therapeutic applications. *PLoS One* (2012) **7**:e37113. doi:10.1371/journal.pone.0037113

26. Wang R, Ding Y, Liu R, Xiang L, Du L. Pomegranate: constituents, bioactivities and pharmacokinetics. *Fruits Veg Cereal Sci Biotech* (2010) **4**:77–87.

27. Medjakovic S, Jungbauer A. Pomegranate: a fruit that ameliorates metabolic syndrome. *Food Funct* (2013) **4**:19–39. doi:10.1039/c2fo30034f

28. Lee KS, Lee BS, Semnani S, Avanesian A, Um CY, Jeon HJ, et al. Curcumin extends life span, improves health span, and modulates the expression of age-associated aging genes

in *Drosophila melanogaster*. *Rejuvenation Res* (2010) **13**:561–70. doi:10.1089/rej.2010.1031

29. Apidianakis Y, Rahme LG. *Drosophila melanogaster* as a model host for studying *Pseudomonas aeruginosa* infection. *Nat Protoc* (2009) **4**:1285–94. doi:10.1038/nprot.2009.124

30. Grandison RC, Piper MD, Partridge L. Amino acid imbalances explains extension of lifespan by dietary restriction in *Drosophila*. *Nature* (2010) **462**:1061–4. doi:10.1038/nature08619

31. Bass TM, Weinkove D, Houthoofd K, Gems D, Partridge L. Effects of resveratrol on lifespan in *Drosophila melanogaster* and *Caenorhabditis elegans*. *Mech Ageing Dev* (2007) **128**:546–52. doi:10.1016/j.mad.2007.07.007

32. Alarco AM, Marcil A, Chen J, Suter B, Thomas D, Whiteway M. Immune-deficient *Drosophila melanogaster*: a model for the innate immune response to human fungal pathogens. *J Immunol* (2004) **172**:5622–8. doi:10.4049/jimmunol.172.9.5622

33. Glittenberg MT, Silas S, MacCallum DM, Gow NA, Ligoxygakis P. Wild-type *Drosophila melanogaster* as an alternative model system for investigating the pathogenicity of *Candida albicans*. *Dis Model Mech* (2011) **4**:504–14. doi:10.1242/dmm.006619

34. Kendig H, Browning CJ, Thomas SA, Wells Y. Health, lifestyle and gender influences on aging well: an Australian longitudinal analysis to guide health promotion. *Front Public Health* (2014) **2**:70. doi:10.3389/fpubh.2014.00070

35. Quinlan J, Tu MT, Langlois EV, Kapoor M, Ziegler D, Fahmi H, et al. Protocol for a systematic review of the association

between chronic stress during the life course and telomere length. *Syst Rev* (2014) **3**:40. doi:10.1186/2046-4053-3-40

36. Lagouge M, Larsson NG. The role of mitochondrial DNA mutations and free radicals in disease and ageing. *J Intern Med* (2013) **273**:529–43. doi:10.1111/joim.12055

37. Khurana S, Venkataraman K, Hollingsworth A, Piche M, Tai TC. Polyphenols: benefits to the cardiovascular system in health and in aging. *Nutrients* (2013)**5**:3779–827. doi:10.3390/nu5103779

38. Chamilos G, Lionakis MS, Lewis RE, Lopez-Ribot JL, Saville SP, Albert ND, et al. *Drosophila melanogaster* as a facile model for large-scale studies of virulence mechanisms and antifungal drug efficacy in *Candida* species. *J Infect Dis* (2006) **193**:1014–22. doi:10.1086/500950

39. Montecino-Rodriguez E, Berent-Maoz B, Dorshkind K. Causes, consequences and reversal of immune system in aging. *J Clin Invest* (2013) **123**:958–65. doi:10.1172/JCI64096

40. Jones MA, Grotewiel M. *Drosophila* as a model for age-related impairment in locomotor and other behaviors. *Exp Gerontol* (2011) **46**:320–5. doi:10.1016/j.exger.2010.08.012

41. Kaeberlein M. Resveratrol and rapamycin: are they anti-aging drugs? *Bioessays* (2010) **32**:96–9. doi:10.1002/bies.200900171

42. World Health Organization and US National Institute of Aging. *Global Health and Aging*. Washington, DC (2011). NIH Publication No. 11-7737. Available from: http://www.who.int/ageing/publications/global_health.pdf?ua=1

8
CHAPTER

Mechanism of Breast Cancer Preventive Action of Pomegranate: Disruption of Estrogen Receptor and Wnt/β-Catenin Signaling Pathways

Animesh Mandal [1] and Anupam Bishayee [2,*]

[1]*Department of Pharmaceutical Sciences, College of Pharmacy, Northeast Ohio Medical University, Rootstown, OH 44272, USA*

[2]*Department of Pharmaceutical Sciences, College of Pharmacy, Larkin Health Sciences Institute, 18301 N. Miami Avenue, Miami, FL 33169, USA*

ABSTRACT

A pomegranate emulsion (PE), containing various bioactive phytochemicals, has recently been found to exert substantial chemopreventive effect against 7,12-dimethylbenz(*a*)anthracene (DMBA)-induced mammary tumorigenesis in rats via antiproliferative and proapoptotic actions. Nevertheless, the underlying mechanisms of action are not completely understood. The present study was designed to investigate the effects of PE treatment on intratumor expression of estrogen receptor (ER)-α, ER-β,β-catenin and cyclin D1 during DMBA rat mammary carcinogenesis. Mammary tumor sections were harvested from a chemopreventive study in which PE (0.2, 1.0 and 5.0 g/kg) exhibited inhibition of mammary tumorigenesis in a dose-response manner. The expressions of ER-α, ER-β, β-catenin and cyclin D1 were analyzed by immunohistochemical techniques. PE downregulated the expression of intratumor ER-α and ER-β and lowered ER-α:ER-β ratio. PE also decreased the expression, cytoplasmic accumulation, and nuclear translocation of β-catenin, an essential transcriptional cofactor for Wnt signaling. Moreover, PE suppressed the expression of cell growth regulatory protein cyclin D1, which is a downstream target for both ER and Wnt signaling. Our current results in conjunction with our previous findings indicate that concurrent disruption of ER and Wnt/β-catenin signaling pathways possibly contributes to antiproliferative and proapoptotic effects involved in PE-mediated chemoprevention of DMBA-inflicted rat mammary tumorigenesis.

Keywords: mammary tumorigenesis; DMBA; pomegranate (*Punica granatum* L.); estrogen receptors; β-catenin; cyclin D1

1. INTRODUCTION

Breast cancer is the most common cancer among women and the second leading cause of cancer-associated deaths in humans worldwide. According to GLOBOCAN 2012 [1], 522,000 women died of breast cancer in both developed and developing countries in 2012. In 2015, approximately 232,000 new breast cancer cases and about 40,000 deaths are estimated to occur in women in the United States [2]. Interestingly, breast cancer also occurs in men with very less frequency compared to women [3]. Genetic risk factors, including mutations on breast cancer susceptibility gene 1 (*BRCA1*) and *BRCA2*, are responsible for about 5%–10% of all breast cancer incidences [4,5]. There are several acquired risk factors for breast cancer which include early onset of menstruation, not having children, delayed birth of a first child, short duration of breast feeding, late menopause, use of hormone replacement therapy, aging, obesity, diabetes, alcohol consumption and circadian disruption [6,7,8,9,10,11,12,13].

Current treatment options for early-stage breast cancer include surgical resection, radiotherapy and adjuvant chemotherapy and/or hormone therapy. Although there are a large numbers of chemotherapeutic drugs, development of resistance and severe adverse side effects represent two serious challenges in the management of breast cancer using these drugs. These facts underscore the importance of developing novel drugs which are more effective and less toxic. Another prudent approach could be chemoprevention which represents prevention of breast cancer occurrence through dietary means and/or use of pharmacological and natural agents [6,7,14,15]. Various epidemiological studies have shown that consumption of high levels of fruits, vegetables and beverages reduce the risk of breast cancer development and recurrence as well as increase the survival rate of patients with this cancer [16,17,18,19,20,21]. Numerous bioactive phytochemicals present in dietary and non-dietary agents have been found to kill breast tumor cells *in vitro* and suppress the development of mammary tumors or retard the growth of existing tumors *in vivo* through modulation of proliferation, differentiation, apoptosis, oxidative stress, inflammation, neovascularization, and several important cell signaling pathways [22,23,24,25,26,27,28,29]. Moreover, several clinical intervention trials investigated the potential efficacy of various dietary supplements and natural products in breast cancer prevention and treatment [30,31].

Pomegranate (*Punica granatum*, L.), an ancient, mystical and highly distinctive fruit, is widely consumed in various parts of the world. Pomegranate fruit has been gaining widespread popularity as a functional food and nutraceutical source due to published reports on potential health benefits, including prevention and/or treatment of oncologic diseases, cardiovascular and neurological disorders, inflammation, ulcer, arthritis, microbial infection, obesity, diabetes, acquired immune deficiency syndrome, and male infertility [32,33,34,35]. Pomegranate fruit represents a rich reservoir of phytochemicals, including polyphenols, hydrolysable tannins, (punicalagin, ellagic acid, gallic acid and gallagic acid), fatty acid (punicic acid) and anthocyanins (delphinidin, cyaniding and pelargonidin). Various preclinical and clinical studies demonstrate chemopreventive and therapeutic effects of pomegranate-derived substances, such as juice, extracts and phytochemicals, against prostate, colon, lung, and skin cancer [36,37]. Based on numerous *in vitro* studies, several pomegranate products and phytoconstituents exhibited cytotoxic, antiproliferative, proapoptotic, antiangiogenic, anti-invasive, and antimetastatic effects against estrogen receptor-positive and -negative breast carcinoma cells [38,39,40,41,42,43,44,45,46,47,48]. Pomegranate seed oil and fermented juice concentrate suppressed 7,12-dimethyl benz(*a*)anthracene (DMBA)-induced precancerous mammary gland lesions *ex vivo* [49] and pomegranate extract inhibited the growth of xenografted BT-474 tumors *in vivo* [45].

Recently, we have reported for the first time that a pomegranate formulation (emulsion) containing most bioactive phytochemicals present in the whole fruit affords a remarkable chemopreventive effect against DMBA-induced mammary tumorigenesis in rats [50]. In this study, pomegranate emulsion (PE) reduced the incidence, total burden and average weight of mammary tumors in DMBA-initiated rats with a concurrent inhibition of cell proliferation, induction of apoptosis, upregulation of proapoptotic protein Bax, and downregulation of antiapoptotic protein Bcl-2 in mammary tumors [50]. Since estrogen receptors (ERs) are involved in mammary cell proliferation [51,52] and DMBA-inflicted rat mammary tumors express ERs [53], we hypothesize that PE-mediated inhibition of mammary tumor cell proliferation could be attained via interference with the expressions of ERs. Moreover, since upregulation of Wnt/β-catenin signaling, which plays a pivotal function in regulation of cell proliferation and apoptosis, has been observed in DMBA-induced mammary tumors in rats [54], it is conceivable that PE could impart antiproliferative and proapoptotic effects through disruption Wnt/β-catenin signaling and thereby blocking expression of downstream genes responsible for promotion of cell proliferation and suppression of apoptotic cell death during rat mammary carcinogenesis. Accordingly, the present study was conducted to extend our previous work [50] to investigate the effects of PE treatment on ER and Wnt/β-catenin signaling as well as expression of cyclin D1, a downstream target for both ER and Wnt signaling, during DMBA-initiated rat mammary tumorigenesis.

2. RESULTS

2.1. PE Suppresses Elevated ER-α and ER-β Expressions during DMBA-Induced Mammary Tumorigenesis

Since ER status is an important classifier of breast cancer, intratumor expressions of ER-α and ER-β in DMBA-initiated rats in the presence or absence of PE treatment were investigated using immunohistochemical techniques. The protein expression of ER-α and ER-β was detected chiefly in the nuclei of epithelial cells. The frequency and intensity of ER-α-immunopositive cells were very high in tumor sections harvested from DMBA control animals (Figure 1A). PE, at 0.2 g/kg, did not alter the expression pattern of ER-α in tumors from DMBA-initiated rats (Figure 1B). On the contrary, a dose-responsive decrease in the expression of ER-α was noticed in tumor sections harvested from animals treated with 1.0 g/kg (Figure 1C) or 5.0 g/kg (Figure 1D) of PE compared to DMBA control. The quantitative analysis reveals immunopositivity for nearly 25% of mammary tumors cells in DMBA control rats (Figure 2A). A significant ($p < 0.001$) reduction in the percentage of ER-α-positive tumor cells in rats treated with 1.0 g/kg or 5.0 g/kg of PE was noticed compared to DMBA control (Figure 2A). Like ER-α, an ample expression of ER-β was observed

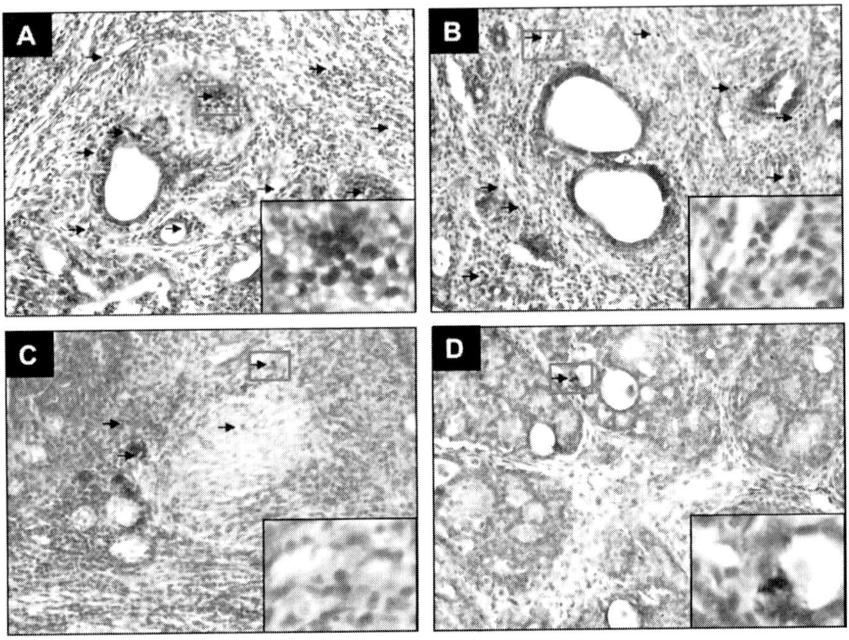

Figure 1. Effect of PE on expression of ER-α during DMBA-induced rat mammary gland tumorigenesis. The rats were treated with various oral doses of PE (three times a week) 2 weeks prior to and 16 weeks after DMBA administration. All animals were sacrificed 16 weeks post-DMBA treatment. The mammary tumors were subjected to immunohistochemical analysis using anti-ER-α antibody. Arrows indicate immunohistochemical staining of ER-α (magnification: ×200). The nuclear expression of ER-α in the designated area marked by red box is shown as an inset (magnification: ×1000) for each treatment group. Various treatment groups are: (**A**) DMBA control; (**B**) PE at 0.2 g/kg body weight plus DMBA; (**C**) PE at 1.0 g/kg body weight plus DMBA; and (**D**) PE at 5.0 g/kg body weight plus DMBA.

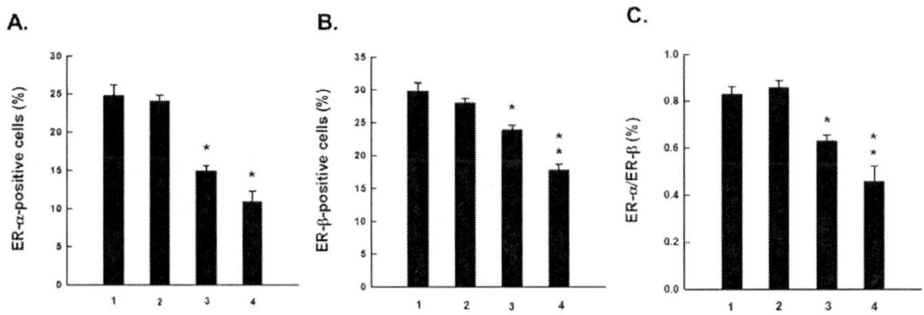

Figure 2. Quantitative analysis of (**A**) ER-α-immunopositive cells; (**B**) ER-β-immunopositive cells and (**C**) ER-α/ER-β ratio during DMBA mammary carcinogenesis in rats in the presence or absence of PE treatment (0.2, 1.0 or 5.0 g/kg). Results are based on 1000 cells per animal and 4 animals per group. Various experimental groups are: (1) DMBA control; (2) PE (0.2 g/kg) + DMBA; (3) PE (1.0 g/kg) + DMBA; and (4) PE (5.0 g/kg) + DMBA. Each bar represents the mean ± SEM ($n = 4$). (**A**) * $p < 0.001$; (**B**) * $p < 0.01$ and ** $p < 0.001$ and (**C**) *$p < 0.05$ and ** $p < 0.001$ as compared to DMBA control.

in tumor samples of rats exposed to DMBA alone (Figure 3A). Although the expression of ER-β was not altered by PE at 0.2 g/kg (Figure 3B), a dose of 1.0 (Figure 3C) or 5.0 g/kg (Figure 3D) displayed considerable attenuation of ER-β immunopositivity. The quantitative analyses of immunopositive cells indicated a significant ($p < 0.01$ and $p < 0.001$) reduction in ER-β-positive cells (Figure 2B) in tumor samples from rats that received 1.0 and 5.0 g/kg PE compared to DMBA control, respectively. These doses of PE attenuated the ratio of ER-α to ER-β in a statistically significant ($p < 0.05$ or 0.001) manner.

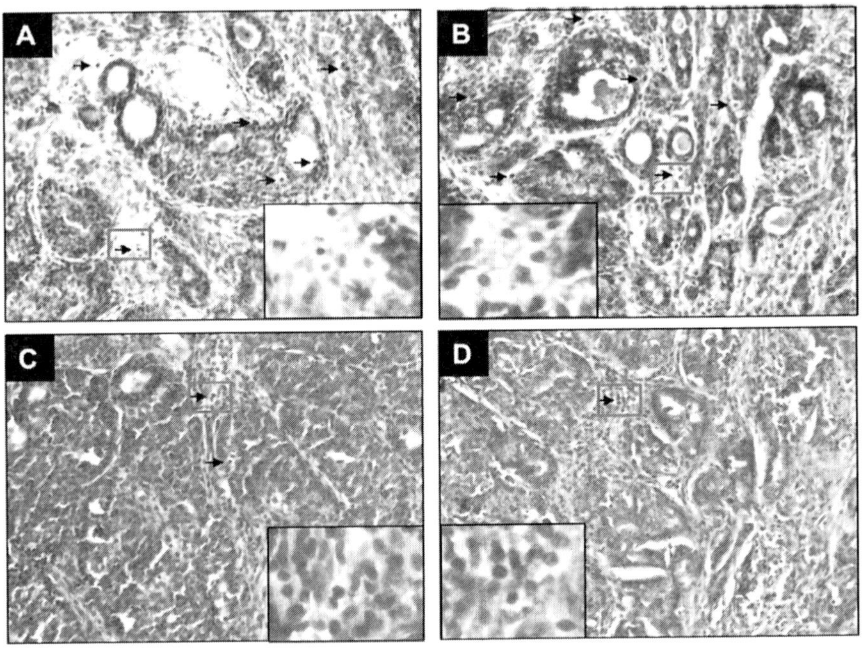

Figure 3. Effect of PE on expression of ER-β during DMBA-induced rat mammary gland tumorigenesis. The rats were treated with various oral doses of PE (three times a week) 2 weeks prior to and 16 weeks after DMBA administration. All animals were sacrificed 16 weeks post-DMBA treatment. The mammary tumors were subjected to immunohistochemical analysis using anti-ER-β antibody. Arrows indicate immunohistochemical staining of ER-β (magnification: ×200). The nuclear expression of ER-β in the designated area marked by red box is shown as an inset (magnification: ×1000) for each treatment group. Various treatment groups are: (**A**) DMBA control; (**B**) PE at 0.2 g/kg body weight plus DMBA; (**C**) PE at 1.0 g/kg body weight plus DMBA; and (**D**) PE at 5.0 g/kg body weight plus DMBA.

2.2. PE Interferes with Activated β-Catenin Signaling during Mammary Tumorigenesis Induced by DMBA

As depicted in Figure 4A, the immunohistochemical profile indicates alteration in nuclear and cytosolic expressions of β-catenin in tumor sections harvested from several animal groups exposed to DMBA. Substantial expression of both nuclear and cytosolic β-catenin-positive cells was observed in rats subjected to DMBA mammary carcinogenesis (Figure 4A-a). The rats which had oral PE at 0.2 g/kg in addition to DMBA showed minimal changes in the expression of nuclear as well as cytosolic expression of β-catenin compared to DMBA control (Figure 4A-b). A considerable decrease in the expression of β-catenin in both nucleus and cytoplasm was achieved by PE treatment at a dose of either 1.0 (Figure 4A-c) or 5.0 g/kg (Figure 4A-d). The accompanying quantitative analysis (Figure 4B,C) confirms our immunohistochemical results, depicting a significant ($p < 0.001$) decrease in nuclear and cytosolic β-catenin expression in rats treated with PE at 1.0 or 5.0 g/kg plus DMBA compared to DMBA alone.

2.3. PE Downregulates Cyclin D1 Expression during DMBA-Induced Mammary Carcinogenesis

Figure 5 illustrates immunohistochemical results showing expression of cell cycle specific gene cyclin D1 in mammary tumor samples from various experimental animal groups. Cyclin D1 was found to be expressed predominantly in the nuclei of tumor cells in DMBA control animals (Figure 5A-a). A marginal alteration in the expression of cyclin D1 was observed in the group that received 0.2 g/kg PE plus DMBA compared to DMBA control (Figure 5A-b). On the other hand, tumor sections from rats that had PE at 1.0 (Figure 5A-c) and 5.0 g/kg (Figure 5A-d) exhibited moderate and substantial reduction in the expression of cyclin D1, respectively. The quantitative analysis indicates a marginal increase in the percentage of cyclin D1-immunopositive cells in the

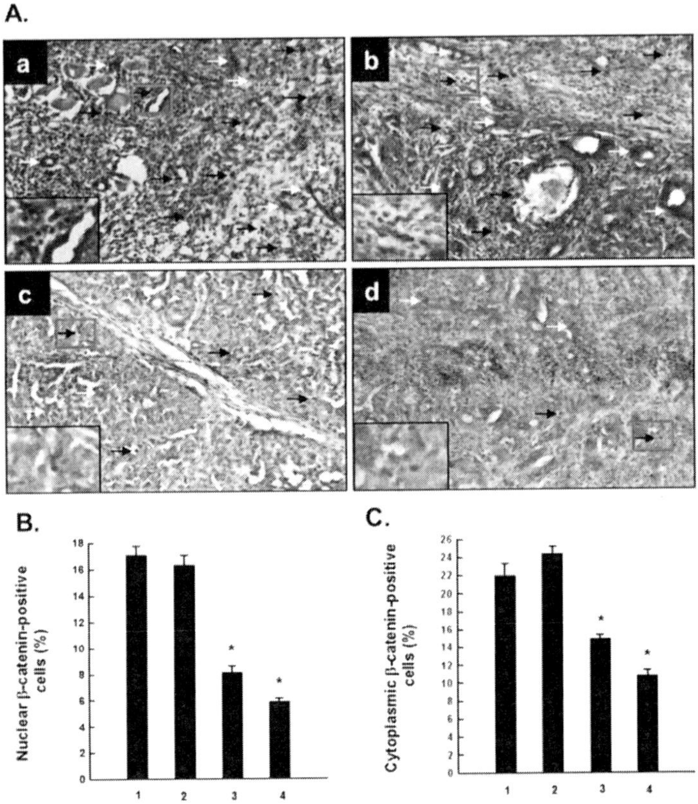

Figure 4. Effect of PE on expression β-catenin during DMBA-induced mammary carcinogenesis in rats. (**A**) Immunohistochemical detection of β-catenin in several experimental rat groups. The rats were treated with various oral doses of PE (three times a week) 2 weeks prior to and 16 weeks after DMBA administration. All animals were sacrificed 16 weeks post-DMBA treatment. The mammary tumors were subjected to immunohistochemical analysis using anti-β-catenin antibody. Representative immunohistochemical localization of β-catenin in nucleus (black arrows) and cytosol (white arrows) is depicted (magnification: ×200). The nuclear expression of β-catenin in the designated area marked by red box is shown as an inset (magnification: ×1000) for each treatment group. Various treatment groups are: (a) DMBA control; (b) PE at 0.2 g/kg body weight plus DMBA; (c) PE at 1.0 g/kg body weight plus DMBA; and (d) PE at 5.0 g/kg body weight plus DMBA; (**B**) Quantitative analysis of nuclear and (**C**) cytoplasmic β-catenin-immunopositive cells in rat mammary tumors induced by DMBA in the presence or absence of PE treatment (0.2, 1.0 or 5.0 g/kg). Results are based on 1000 cells per animal and 4 animals per group. Each bar represents the mean ± SEM (n = 4). * $p < 0.001$ as compared to DMBA control. Various experimental groups are: (1) DMBA control; (2) PE (0.2 g/kg) + DMBA; (3) PE (1.0 g/kg) + DMBA; and (4) PE (5.0 g/kg) + DMBA.

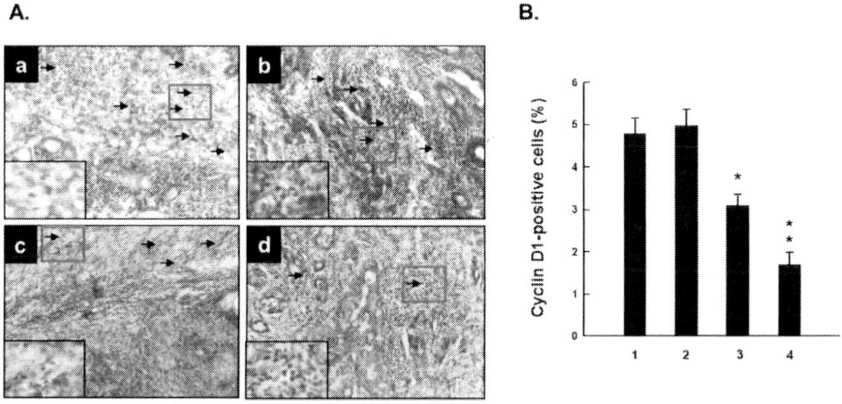

Figure 5. Effect of PE on expression of cyclin D1 during DMBA-inflicted mammary gland tumorigenesis in rats. (**A**) Immunohistochemical detection of cyclin D1 in several experimental rat groups. The rats were treated with various oral doses of PE (three times a week) 2 weeks prior to and 16 weeks after DMBA administration. All animals were sacrificed 16 weeks post-DMBA treatment. The mammary tumors were subjected to immunohistochemical analysis using anti-cyclin D1 antibody. Arrows indicate immunohistochemical localization of cyclin D1 in nucleus (magnification: ×200). The nuclear expression of cyclin D1 in the designated area marked by red box is shown as an inset (magnification: ×1000) for each treatment group. Various treatment groups are: (a) DMBA control; (b) PE at 0.2 g/kg body weight plus DMBA; (c) PE at 1.0 g/kg body weight plus DMBA; and (d) PE at 5.0 g/kg body weight plus DMBA; (**B**) Quantitative analysis of cyclin D1-immunopositive cells in rat mammary tumors induced by DMBA in the presence or absence of PE treatment (0.2, 1.0 or 5.0 g/kg). Results are based on 1000 cells per animal and 4 animals per group. Each bar represents the mean ± SEM (n = 4). * $p < 0.05$ and ** $p < 0.001$ as compared to DMBA control. Various experimental groups are: (1) DMBA control; (2) PE (0.2 g/kg) + DMBA; (3) PE (1.0 g/kg) + DMBA; and (4) PE (5.0 g/kg) + DMBA.

group treated with 0.2 g/kg PE compared to DMBA control (Figure 5B). However, a significant ($p < 0.05$ or 0.001) decrement in the percentage of cyclin D1-positive cells was achieved in DMBA-initiated rats treated with PE at 1.0 or 5.0 g/kg in comparison with DMBA control, respectively (Figure 5B).

3. DISCUSSION

Elevated lifetime exposure to endogenous or exogenous estrogen has been accepted as the single most important risk factor in the development of breast cancer [55]. Estrogen activation of ERs triggers specific signaling pathways responsible for mammary cell proliferation and differentiation and plays a pivotal role in the development of healthy mammary glands. On the other hand, altered ER signaling is associated with abnormal cell proliferation as well as initiation and progression of breast cancer [51,52]. Emerging evidence suggest that 70% of breast cancers express ERs [56]. It is well known that estrogen binds to ER-α and ER-β and activates transcription of estrogen-responsive genes, resulting in accelerated tumor cell proliferation [57,58]. Accordingly, the modulation of ER-α and ER-β may be a central mechanism to suppress mammary gland carcinogenesis [59,60]. Use of antiestrogenic drugs capable of competing with estrogen for binding to ERs is known to be effective in impeding estrogen-dependent mammary tumor growth as well as preventing the occurrence of breast tumor [14,61,62].

In our study, we have evaluated the expression of ER-α and ER-β in DMBA-induced mammary tumors in rats subjected to PE treatment. Our results indicate considerable expression of ER-α and ER-β in mammary tumors harvested from DMBA control animals. The oral administration of PE to rats before and after DMBA treatment decreased both ER-α and ER-β protein expression in mammary tumors. PE may reduce ER gene transcription, translation or induce epigenetic modifications.

However, additional studies are warranted to confirm the exact mechanism of reduced levels of ER-α and ER-β in DMBA-induced tumors in PE treatment groups. Another salient feature of our study is PE-mediated decrement in the ratio of ER-α to ER-β. An earlier report documents an increase in the ERα:ER-β in human breast tumorigenesis [59]. Moreover, an elevated ERα:ER-β correlates with higher level of cell proliferation in preinvasive human mammary tumors [63]. Phytoconstituents present in PE may compromise the responsiveness to endogenous estrogen by diminishing the expression of ERs, resulting reduced availability of nuclear receptor sites for estrogen binding during experimental mammary carcinogenesis. The overall effect could be a significant reduction in intratumor proliferation by a reduced expression of proliferating cell nuclear antigen as we reported recently [50]. Furthermore, the prior exposure of the mammary glands to PE may compromise the ability of ER-positive cells to respond to DMBA challenge. Collectively, all these attributes could be responsible for subsequent development of fewer proliferating mammary tumors as we have observed in our previous study [50]. Interestingly, our results are in line with a previous report showing that a methanolic extract of pericarp (peel) of pomegranate inhibited the binding of estradiol to ER, downregulated ER-α gene, and suppressed the growth and proliferation of ER-positive MCF-7 breast cancer cells [47].

It is likely that estrogens impart their oncogenic effects through regulation of ER-dependent as well as ER-independent pathways. Therefore, the net mammary tumor inhibitory effect of PE in DMBA-initiated, estrogen-dependent mammary carcinogenesis model could be attributed to PE-mediated disruption of ER signaling as well as non-ER mediated mechanisms. As a matter of fact, several pomegranate-derived products and phytochemicals inhibited the growth of ER-negative MBA-MD-231 breast cancer cells [38,45].

The canonical Wnt or Wnt/β-catenin signaling pathway plays an important role in mammary gland development as well as tumorigenesis due to its involvement in signal transduction, cellular adhesion, and regulation of cell-context-specific gene expression [64,65]. In an absence of a Wnt signal, β-catenin is linked to a multiprotein complex, consists of Axin, adenomatous polyposis coli, casein kinase-1α, and glycogen synthase kinase, which facilitates its phosphorylation, ubiquitination and degradation by proteasome [66]. In the presence of a Wnt signal, the destruction complex is inactivated, which leads to stabilization and accumulation of β-catenin in the cytoplasm. Consequently, β-catenin translocates to the nucleus and binds to T-cell factor/lymphoid-enhancer factor, resulting in activation of transcription of various target genes, such as c-myc, cyclin D1, matrix metalloproteinase 7, and vascular endothelial growth factor, which are implicated in mammary gland carcinogenesis [67]. A significant accumulation of β-catenin in the nucleus and/or cytoplasm has been observed in human breast carcinoma samples and is believed to be associated with poor prognosis [68,69]. Interestingly, upregulation of nuclear and/or cytoplasmic β-catenin in human breast cancer samples has been correlated with the expression of its target gene - cyclin D1 [70]. Additionally, an elevated accumulation of cytosolic and nuclear β-catenin has been observed in ductal carcinoma *in situ* and basal-like *in situ* breast tumors, indicating that activation of Wnt/β-catenin pathway may be an early event in human breast cancer [68,71]. Consistent with clinical situations, an increase in total and nuclear β-catenin protein has been found in DMBA-induced mammary tumors in mice, showing stimulation of this oncogenic signaling pathway [72]. Furthermore, a sequential elevation of β-catenin level in mammary tissues have been found in rats subjected to DMBA mammary carcinogenesis [54]. In our study, a prominent nuclear and cytoplasmic β-catenin expression in the tumors harvested from DMBA control rats confirms the activation of Wnt/β-catenin pathway at an early stage of chemically induced mammary carcinogenesis in rats. Our results also revealed that PE treatment caused abrogation of Wnt/β-catenin signaling marked by reduced expression of nuclear and cytosolic β-catenin. Since down-modulation of Wnt/β-catenin signaling has been associated with inhibition of cellular proliferation and induction of apoptosis [73], suppression of constitutive activation of Wnt/β-catenin signaling appears to be a possible mechanism of PE-mediated inhibition of cell proliferation and escalation of apoptosis in DMBA-initiated mammary tumorigenesis in rats as we reported earlier [50]. Previously, we have observed that PE exerted similar antiproliferative and proapoptotic effects through modulation of Wnt/β-catenin signaling during chemically induced rat liver carcinogenesis [74]. An ellagitannin-rich pomegranate extract, ellagic acid, and colonic metabolite urolithin have been shown to inhibit canonical Wnt signaling pathway in human 293T cell line [75]. Similarly, a standardized pomegranate extract has been found to modulate several components of Wnt//β-catenin signaling during chemical rat colon carcinogenesis [76].

Cyclin D1, a cofactor for ER action, contributes to mammary tumorigenesis, by the regulation of proliferation and differentiation [77,78]. The mRNA and protein levels of cyclin D1 have been found to be upregulated in more than 50% of the breast cancers and cyclin D1 represents one of the most commonly overexpressed proteins in this cancer [79]. Cyclin D1, a known target for ER, has been reported to be overexpressed favorably in ER-positive breast cancer [79]. In our current study, a substantial expression of intratumor cyclin D1 in DMBA control rats underscores the fundamental role of this cell cycle regulatory protein in DMBA-initiated rat mammary tumorigenesis, supporting previous observations [80,81,82]. A radical reduction of cyclin D1 protein expression due to PE treatment has been observed in our study which indicates that reversal of DMBA-induced dysregulation of a critical cell cycle checkpoint may be one of the possible molecular mechanisms of PE-mediated suppression of mammary tumorigenesis. Our findings also suggest that cyclin D1 may be a potential target of pomegranate bioactive constituents for the chemoprevention of breast cancer. Since cyclin D1 is a β-catenin-regulated gene [83] and PE downregulated cyclin D1 in the same manner as β-catenin, our data confirm interference of Wnt/β-catenin signaling by PE during chemical rat mammary carcinogenesis. Previously, we [74] and other investigators [76] reported pomegranate extract-mediated inhibition of cyclin D1 during experimentally induced rat hepatocarcinogenesis and colon carcinogenesis, respectively.

The identification of specific bioactive phytochemicals of PE responsible for the observed effects in terms with various end-point biomarkers are not evident at this time and requires additional studies. Several pomegranate phytochemicals present in the PE used in this study exhibited synergistic effects in suppressing growth of tumor cells [84,85]. Emerging evidence suggests that plant phytochemicals exert cancer preventive and anticancer effects when they are used in combination rather than individually [86,87]. Accordingly, it is likely that pomegranate phytochemicals may confer the observed activities via promotion of multifactorial effects utilizing chemical synergy.

4. MATERIALS AND METHODS

4.1. Materials

PE, a proprietary formulation consists of pomegranate aqueous phase extract and seed oil, was purchased from Rimonest Ltd. (Haifa, Israel). We have previously published a detailed description of the preparation of this product [88]. The chemical analyses of this formulation indicated the presence of mixed octadecatrienoic acids, sterols and steroids, especially 17-α-estradiol, and tocol and γ-tocopherol in the lipid phase and caffeic acid, corilagin, ellagic acid, ferulic acid, gallic acid, 5-hydroxymethylfurfural, protocatechuic acid, punicalagins (A and

B) and *trans-p*-coumaric acid in the aqueous phase [88]. The chemical carcinogen DMBA was procured from Sigma-Aldrich (St. Louis, MO, USA). Primary antibodies, such as rabbit polyclonal ER-α (sc-542), ER-β (sc-8974), β-catenin (sc-7199), and cyclin D1 (sc-753) as well as rabbit ABC staining system (sc-2018) were obtained from Santa Cruz Biotechnology (Santa Cruz, CA, USA).

4.2. Animals and Experimental Design

Mammary tumor samples for this work were harvested from our previously completed chemopreventive study [50] based on an animal protocol approved by the Institutional Animal Care and Use Committee of Northeast Ohio Medical University (Rootstown, OH, USA). In brief, female Sprague-Dawley rats (Harlan Laboratories, Indianapolis, IN, USA) were divided into six groups. Two groups (groups A and B) were maintained on a basal diet (LabDiet, St. Louis, MO, USA) *ad libitum*, whereas the remaining four groups (groups C, D, E and F) were fed with PE *per os* (p.o.) three times a week (Monday-Wednesday-Friday) in addition to have access to the aforementioned basal diet. Three doses of PE were used for this work: 0.2 g/kg (group C), 1.0 g/kg (group D) and 5.0 g/kg (groups E and F). After the aforementioned treatment period of 2 weeks, mammary tumorigenesis was initiated in all animals belonging to groups B, C, D and E by a single oral administration (p.o.) of DMBA at a dose of 50 mg/kg body weight. Oral treatment of rats with PE in groups C, D, E and F were continued for another 16 consecutive weeks following DMBA administration, *i.e.*, a total period of 18 weeks. After this period, all animals were sacrificed and mammary tumor samples from various DMBA-exposed animals (groups B, C, D and E) were harvested and fixed in 4% paraformaldehyde for immunohistochemical analysis.

4.3. Immunohistochemical Analysis

Serial tumor sections, approximately 15-μm thick, were cut using a microtome and stored at −80 °C freezer. For immunohistochemical studies, we used similar regions/locations of tumor mass from each group to prepare tissue sections. We also tried to select tumors with similar size to the extent possible. However, most of the tumors excised from animals treated with medium or high dose of PE had smaller sizes as reported earlier [50]. Intratumor expressions of ER-α, ER-β,β-catenin and cyclin D1 were determined by immunohistochemistry. In short, frozen tissue sections were thawed, air dried for 30 min, and subjected to antigen retrieval by immersing in sodium citrate buffer (10 mM, pH 6.0) at a temperature up to 80 °C for 10 min. Endogenous peroxidases were blocked by 1% H_2O_2 (5 min) followed by washing the sections with phosphate-buffered saline (PBS) for 5 min. Tissue sections were then treated with blocking solution for 1 h followed by washing with PBS and incubation overnight (at 4 °C) with primary antibodies (ER-α, ER-β,β-catenin or cyclin D) at a dilution of 1:100. After several washes, tissue sections were treated with horseradish peroxidase-conjugated goat anti-rabbit secondary antibody (1:200) for 30 min at room temperature and then with 3,3'-diaminobenzidine tetrahydrocholoride solution to observe the antigen-antibody complexes. Finally, tumor sections were slightly counterstained with Gill's hematoxylin solution, air dried, and mounted using DPX (Electron Microscopy Sciences, Hatfield, PA, USA). Various sections of slides were chosen randomly and visualized under a light microscope (BX43, Olympus, Center Valley, PA, USA) which was used to capture representative images. An immunopositive tumor cell expressing an antigen was identified based on brown staining. At least 1000 tumor cells/animal were analyzed. Quantitative results were expressed as percentage of immunopositive cells.

4.4. Statistical Analysis

Results are presented as mean ± standard error of the mean (SEM). Significant differences among various treatment groups were determined by one-way ANOVA. *Post hoc* analysis was performed by the Student-Neuman-Keuls test. A *p* value less than 0.05 was considered to be statistically significant. All analyses were performed using commercial software SigmaStat 3.1 (Systat Software, Inc., San Jose, CA, USA).

5. CONCLUSIONS

Based on results presented here, we conclude that PE abrogates the expression of ER-α and ER-β during DMBA-inflicted mammary tumorigenesis in rats. PE also averts cytosolic stabilization, accumulation, and nuclear translocation of β-catenin, an essential transcriptional cofactor for Wnt/β-catenin signaling. Moreover, PE downregulates the expression of cyclin D1, a downstream target for both ER and Wnt signaling pathways. Based on our earlier study, we have demonstrated a striking mammary tumor-inhibitory effect of PE with concomitant antiproliferative and apoptosis-inducing activities under the same experimental conditions [50]. Hence, the current results together with our previous findings indicate that concurrent disruption of ER and Wnt/β-catenin signaling cascades possibly contributes to antiproliferative and proapoptotic effects involved in PE-mediated prevention of DMBA-initiated mammary carcinogenesis in rats. These encouraging preclinical results coupled with a safety profile may facilitate the development PE as a chemopreventive drug to reduce the risk of breast cancer.

ACKNOWLEDGMENTS

A portion of this study was conducted at the Northeast Ohio Medical University (Rootstown, OH, USA) supported by a new faculty research start-up grant awarded to A.B.

AUTHOR CONTRIBUTIONS

A.B. conceived and designed the experiments; A.M. performed the experiments; A.B. and A.M. analyzed the data; A.B. contributed reagents/materials/analysis tools; A.B. wrote the paper.

CONFLICTS OF INTEREST

The authors declare no conflict of interest. The founding sponsors had no role in the design of the study; in the collection, analyses, or interpretation of data; in the writing of the manuscript, and in the decision to publish the results.

REFERENCES

1. Ferlay, J.; Soerjomataram, I.; Dikshit, R.; Eser, S.; Mathers, C.; Rebelo, M.; Parkin, D.M.; Forman, D.; Bray, F. Cancer incidence and mortality worldwide: Sources, methods and major patterns in GLOBOCAN 2012. *Int. J. Cancer* **2015**, *136*, E359–E386.

2. Siegel, R.L.; Miller, K.D.; Jemal, A. Cancer statistics, 2015. *CA Cancer J. Clin.* **2015**, *65*, 5–29.

3. Ottini, L. Male breast cancer: A rare disease that might uncover underlying pathways of breast cancer. *Nat. Rev. Cancer* **2014**, *14*, 643.

4. Lillie, S.E.; Brewer, N.T.; O'Neill, S.C.; Morrill, E.F.; Dees, E.C.; Carey, L.A.; Rimer, B.K. Retention and use of breast cancer recurrence risk information from genomic tests: The role of health literacy. *Cancer Epidemiol. Biomark. Prev.* **2007**, *16*, 249–255.

5. Campeau, P.M.; Foulkes, W.D.; Tischkowitz, M.D. Hereditary breast cancer: New genetic developments, new therapeutic avenues. *Hum. Genet.* **2008**, *124*, 31–42.

6. Bozovic-Spasojevic, I.; Azambuja, E.; McCaskill-Stevens, W.; Dinh, P.; Cardoso, F. Chemoprevention for breast cancer. *Cancer Treat. Rev.* **2012**, *38*, 329–339.

7. Files, J.A.; Stan, D.L.; Allen, S.V.; Pruthi, S. Chemoprevention of breast cancer. *Women's Health***2012**, *8*, 635–646.

8. Fagundes, C.P.; Lindgren, M.E.; Shapiro, C.L.; Kiecolt-Glaser, J.K. Child maltreatment and breast cancer survivors: Social support makes a difference for quality of life, fatigue and cancer stress.*Eur. J. Cancer* **2012**, *48*, 728–736.

9. Narod, S.A. Hormone replacement therapy and the risk of breast cancer. *Nat. Rev. Clin. Oncol.***2011**, *8*, 669–676.

10. Park, S.Y.; Kolonel, L.N.; Lim, U.; White, K.K.; Henderson, B.E.; Wilkens, L.R. Alcohol consumption and breast cancer risk among women from five ethnic groups with light to moderate intakes: The Multiethnic Cohort Study. *Int. J. Cancer* **2014**, *134*, 1504–1510.

11. Crujeiras, A.B.; Díaz-Lagares, A.; Carreira, M.C.; Amil, M.; Casanueva, F.F. Oxidative stress associated to dysfunctional adipose tissue: A potential link between obesity, type 2 diabetes mellitus and breast cancer. *Free Radic. Res.* **2013**, *47*, 243–256.

12. Pierobon, M.; Frankenfeld, C.L. Obesity as a risk factor for triple-negative breast cancers: A systematic review and meta-analysis. *Breast Cancer Res. Treat.* **2013**, *137*, 307–314.

13. Jia, Y.; Lu, Y.; Wu, K.; Lin, Q.; Shen, W.; Zhu, M.; Huang, S.; Chen, J. Does night work increase the risk of breast cancer? A systematic review and meta-analysis of epidemiological studies. *Cancer Epidemiol.* **2013**, *37*, 197–206.

14. Cazzaniga, M.; Bonanni, B. Breast cancer chemoprevention: Old and new approaches. *J. Biomed. Biotechnol.* **2012**, *2012*, 985620.

15. Eccles, S.A.; Aboagye, E.O.; Ali, S.; Anderson, A.S.; Armes, J.; Berditchevski, F.; Blaydes, J.P.; Brennan, K.; Brown, N.J.; Bryant, H.E.; *et al.* Critical research gaps and translational priorities for the successful prevention and treatment of breast cancer. *Breast Cancer Res.* **2013**, *15*, R92.

16. Song, J.K.; Bae, J.M. Citrus fruit intake and breast cancer risk: A quantitative systemic review. *J. Breast Cancer* **2013**, *16*, 72–76.

17. Ogunleye, A.A.; Xue, F.; Michels, K.B. Green tea consumption and breast cancer risk or recurrence: A meta-analysis. *Breast Cancer Res. Treat.* **2010**, *119*, 477–484.

18. Mourouti, N.; Panagiotakos, D.B. Soy food consumption and breast cancer. *Maturitas* **2013**, *76*, 118–122.

19. Aune, D.; Chan, D.S.; Vieira, A.R.; Navarro Rosenblatt, D.A.; Vieira, R.; Greenwood, D.C.; Norat, T. Dietary compared with blood concentrations of carotenoids and breast cancer risk: A systematic review and meta-analysis of prospective studies. *Am. J. Clin. Nutr.* **2012**, *96*, 356–373.

20. Thomson, C.A.; Flatt, S.W.; Rock, C.L.; Ritenbaugh, C.; Newman, V.; Pierce, J.P. Increased fruit, vegetable and fiber intake and lower fat intake reported among women previously treated for invasive breast cancer. *J. Am. Diet Assoc.* **2002**, *102*, 801–808.

21. Rock, C.L.; Demark-Wahnefried, W. Nutrition and survival after the diagnosis of breast cancer: A review of the evidence. *J. Clin. Oncol.* **2002**, *20*, 3302–3316.

22. Khan, S.I.; Zhao, J.; Khan, I.A.; Walker, L.A.; Dasmahapatra, A.K. Potential utility of natural products as regulators of breast cancer-associated aromatase promoters. *Reprod. Biol. Endocrinol.***2011**, *9*, 91.

23. Reuben, S.C.; Gopalan, A.; Petit, D.M.; Bishayee, A. Modulation of angiogenesis by dietary phytoconstituents in the prevention and intervention of breast cancer. *Mol. Nutr. Food Res.* **2012**,*56*, 14–29.

24. Sinha, D.; Biswas, J.; Sung, B.; Aggarwal, B.B.; Bishayee, A. Chemopreventive and chemotherapeutic potential of curcumin in breast cancer. *Curr. Drug Targets* **2012**, *13*, 1799–1819.

25. Aiyer, H.S.; Warri, A.M.; Woode, D.R.; Hilakivi-Clarke, L.; Clarke, R. Influence of berry polyphenols on receptor signaling and cell-death pathways: Implications for breast cancer prevention. *J. Agric. Food Chem.* **2012**, *60*, 5693–5708.

26. Vadodkar, A.S.; Suman, S.; Lakshmanaswamy, R.; Damodaran, C. Chemoprevention of breast cancer by dietary compounds. *Anticancer Agents Med. Chem.* **2012**, *12*, 1185–1202.

27. Takemura, H.; Sakakibara, H.; Yamazaki, S.; Shimoi, K. Breast cancer and flavonoids—A role in prevention. *Curr. Pharm. Des.* **2013**, *19*, 6125–6132.

28. Yiannakopoulou, E.C. Effect of green tea catechins on breast carcinogenesis: A systematic review of *in vitro* and *in vivo* experimental studies. *Eur. J. Cancer Prev.* **2014**, *23*, 84–89.

29. Parikh, N.R.; Mandal, A.; Bhatia, D.; Siveen, K.S.; Sethi, G.; Bishayee, A. Oleanane triterpenoids in the prevention and therapy of breast cancer: Current evidence and future perspectives.*Phytochem. Rev.* **2014**, *13*, 793–810.

30. Kado, K.; Forsyth, A.; Patel, P.R.; Schwartz, J.A. Dietary supplements and natural products in breast cancer trials. *Front. Biosci.* **2012**, *4*, 546–567.

31. Siddiqui, J.A.; Singh, A.; Chagtoo, M.; Singh, N.; Godbole, M.M.; Chakravarti, B. Phytochemicals for breast cancer therapy: Current status and future implications. *Curr. Cancer Drug Targets* **2015**,*15*, 116–135.

32. Lansky, E.P.; Newman, R.A. *Punica granatum* (pomegranate) and its potential for prevention and treatment of inflammation and cancer. *J. Ethnopharmacol.* **2007**, *109*, 177–206.

33. Jurenka, J.S. Therapeutic applications of pomegranate (*Punica granatum* L.): A review. *Altern. Med. Rev.* **2008**, *13*, 128–144.

34. Johanningsmeier, S.D.; Harris, G.K. Pomegranate as a functional food and nutraceutical source.*Annu. Rev. Food Sci. Technol.* **2011**, *2*, 181–201.

35. Faria, A.; Calhau, C. The bioactivity of pomegranate: Impact on health and disease. *Crit. Rev. Food Sci. Nutr.* **2011**, *51*, 626–634.

36. Syed, D.N.; Chamcheu, J.C.; Adhami, V.M.; Mukhtar, H. Pomegranate extracts and cancer prevention: molecular and cellular activities. *Anticancer Agents Med. Chem.* **2013**, *13*, 1149–1161.

37. Turrini, E.; Ferruzzi, L.; Fimognari, C. Potential effects of pomegranate polyphenols in cancer prevention and therapy. *Oxid. Med Cell. Longev.* **2015**, *2015*, 938475.

38. Kim, N.D.; Mehta, R.; Yu, W.; Neeman, I.; Livney, T.; Amichay, A.; Poirier, D.; Nicholls, P.; Kirby, A.; Jiang, W.; *et al.* Chemopreventive and adjuvant therapeutic potential of pomegranate (*Punica granatum*) for human breast cancer. *Breast Cancer Res. Treat.* **2002**, *71*, 203–217.

39. Toi, M.; Bando, H.; Ramachandran, C.; Melnick, S.J.; Imai, A.; Fife, R.S.; Carr, R.E.; Oikawa, T.; Lansky, E.P. Preliminary studies on the anti-angiogenic potential of pomegranate fractions *in vitro*and *in vivo*. *Angiogenesis* **2003**, *6*, 121–128.

40. Khan, G.N.; Gorin, M.A.; Rosenthal, D.; Pan, Q.; Bao, L.W.; Wu, Z.F.; Newman, R.A.; Pawlus, A.D.; Yang, P.; Lansky, E.P.; *et al.* Pomegranate fruit extract impairs invasion and motility in human breast cancer. *Integr. Cancer Ther.* **2009**, *8*, 242–253.

41. Adams, L.S.; Zhang, Y.; Seeram, N.P.; Heber, D.; Chen, S. Pomegranate ellagitannin-derived compounds exhibit antiproliferative and antiaromatase activity in breast cancer cells *in vitro*.*Cancer Prev. Res.* **2010**, *3*, 108–113.

42. Dai, Z.; Nair, V.; Khan, M.; Ciolino, H.P. Pomegranate extract inhibits the proliferation and viability of MMTV-Wnt-1

mouse mammary cancer stem cells *in vitro*. *Oncol. Rep.* **2010**, *24*, 1087–1091.

43. Dikmen, M.; Ozturk, N.; Ozturk, Y. The antioxidant potency of Punica granatum L. fruit peel reduces cell proliferation and induces apoptosis on breast cancer. *J. Med. Food* **2011**, *14*, 1638–1646.

44. Joseph, M.M.; Aravind, S.R.; Varghese, S.; Mini, S.; Sreelekha, T.T. Evaluation of antioxidant, antitumor and immunomodulatory properties of polysaccharide isolated from fruit rind of *Punica granatum*. *Mol. Med. Rep.* **2012**, *5*, 489–496.

45. Banerjee, N.; Talcott, S.; Safe, S.; Mertens-Talcott, S.U. Cytotoxicity of pomegranate polyphenolics in breast cancer cells *in vitro* and *vivo*: Potential role of miRNA-27a and miRNA-155 in cell survival and inflammation. *Breast Cancer Res. Treat.* **2012**, *136*, 21–34.

46. Rocha, A.; Wang, L.; Penichet, M.; Martins-Green, M. Pomegranate juice and specific components inhibit cell and molecular processes critical for metastasis of breast cancer. *Breast Cancer Res. Treat.* **2012**, *136*, 647–658.

47. Sreeja, S.; Santhosh Kumar, T.R.; Lakshmi, B.S.; Sreeja, S. Pomegranate extract demonstrate a selective estrogen receptor modulator profile in human tumor cell lines and *in vivo* models of estrogen deprivation. *J. Nutr. Biochem.* **2012**, *23*, 725–732.

48. Shirode, A.B.; Kovvuru, P.; Chittur, S.V.; Henning, S.M.; Heber, D.; Reliene, R. Antiproliferative effects of pomegranate extract in MCF-7 breast cancer cells are associated with reduced DNA repair gene expression and induction of double strand breaks. *Mol. Carcinog.* **2014**, *53*, 458–470.

49. Mehta, R.; Lansk, E.P. Breast cancer chemopreventive properties of pomegranate (*Punica granatum*) fruit extracts in a mouse mammary organ culture. *Eur. J. Cancer Prev.* **2004**, *13*, 345–348.

50. Bishayee, A.; Mandal, A.; Bhattacharyya, P.; Bhatia, D. Pomegranate exerts chemoprevention of experimentally induced mammary tumorigenesis by suppression of cell proliferation and induction of apoptosis. *Nutr. Cancer* **2015**.

51. Matthews, J.; Gustafsson, J.A. Estrogen signaling: A subtle balance between ER alpha and ER beta.*Mol. Interv.* **2003**, *3*, 281–292.

52. Williams, G.P. The role of oestrogen in the pathogenesis of obesity, type 2 diabetes, breast cancer and prostate disease. *Eur. J. Cancer Prev.* **2010**, *19*, 256–271.

53. Russo, J.; Russo, I.H. Experimentally induced mammary tumors in rats. *Breast Cancer Res. Treat.***1996**, *39*, 7–20.

54. Kilańczyk, E.; Gwoździński, K.; Wilczek, E.; Filipek, A. Up-regulation of CacyBP/SIP during rat breast cancer development. *Breast Cancer* **2014**, *21*, 350–357.

55. Yager, J.D.; Davidson, N.E. Estrogen carcinogenesis in breast cancer. *N. Engl. J. Med.* **2006**, *354*, 270–282.

56. Althuis, M.D.; Fergenbaum, J.H.; Garcia-Closas, M.; Brinton, L.A.; Madigan, M.P.; Sherman, M.E. Etiology of hormone receptor-defined breast cancer: A systematic review of the literature. *Cancer Epidemiol. Biomark. Prev.* **2004**, *13*, 1558–1568.

57. Korach, K.S.; Couse, J.F.; Curtis, S.W.; Washburn, T.F.; Lindzey, J.; Kimbro, K.S.; Eddy, E.M.; Migliaccio, S.; Snedeker, S.M.; Lubahn, D.B.; *et al.* Estrogen receptor gene

disruption: Molecular characterization and experimental and clinical phenotypes. *Recent Prog. Horm. Res.* **1996**, *51*, 159–186.

58. Saji, S.; Jensen, E.V.; Nilsson, S.; Rylander, T.; Warner, M.; Gustafsson, J.A. Estrogen receptors alpha and beta in the rodent mammary gland. *Proc. Natl. Acad. Sci. USA* **2000**, *97*, 337–342.

59. Leygue, E.; Dotzlaw, H.; Watson, P.H.; Murphy, L.C. Altered estrogen receptor α and β messenger RNA expression during human breast tumorigenesis. *Cancer Res.* **1998**, *58*, 3197–3201.

60. Lazennce, G.; Bresson, D.; Lucas, A.; Chauveau, C.; Vignon, F. ER beta inhibits proliferation and invasion of breast cancer cells. *Endocrinology* **2001**, *142*, 4120–4130.

61. Jordan, V.C. Chemoprevention of breast cancer with selective oestrogen-receptor modulators. *Nat. Rev. Cancer* **2007**, *7*, 46–53.

62. Cuzick, J.; DeCensi, A.; Arun, B.; Brown, P.H.; Castiglione, M.; Dunn, B.; Forbes, J.F.; Glaus, A.; Howell, A.; von Minckwitz, G.; *et al.* Preventive therapy for breast cancer: A consensus statement.*Lancet Oncol.* **2011**, *12*, 496–503.

63. Roger, P.; Sahla, M.E.; Mäkelä, S.; Gustafsson, J.A.; Baldet, P.; Rochefort, H. Decreased expression of estrogen receptor beta protein in proliferative preinvasive mammary tumors. *Cancer Res.* **2001**,*61*, 2537–2541.

64. Brennan, K.R.; Brown, A.M.C. Wnt proteins in mammary development and cancer. *J. Mammary Gland Biol. Neoplasia* **2004**, *9*, 119–131.

65. Incassati, A.; Chandramouli, A.; Eelkema, R.; Cowin, P. Key signaling nodes in mammary gland development and cancer: β-catenin. *Breast Cancer Res.* **2010**, *12*, 213.

66. Clevers, H. Wnt/β-catenin signaling in development and disease. *Cell* **2006**, *127*, 469–480.

67. Ozaki, S.; Ikeda, S.; Ishizaki, Y.; Kurihara, T.; Tokumoto, N.; Iseki, M.; Arihiro, K.; Kataoka, T.; Okajima, M.; Asahara, T. Alterations and correlations of the components in the Wnt signaling pathway and its target genes in breast cancer. *Oncol. Rep.* **2005**, *14*, 1437–1443.

68. Khramtsov, A.I.; Khramtsova, G.F.; Tretiakova, M.; Huo, D.; Olopade, O.I.; Goss, K.H. Wnt/beta-catenin pathway activation is enriched in basal-like breast cancers and predicts poor outcome.*Am. J. Pathol.* **2010**, *176*, 2911–2920.

69. López-Knowles, E.; Zardawi, S.J.; McNeil, C.M.; Millar, E.K.; Crea, P.; Musgrove, E.A.; Sutherland, R.L.; O'Toole, S.A. Cytoplasmic localization of beta-catenin is a marker of poor outcome in breast cancer patients. *Cancer Epidemiol. Biomarkers Prev.* **2010**, *19*, 301–309.

70. Prasad, C.P.; Gupta, S.D.; Rath, G.; Ralhan, R. Wnt signaling pathway in invasive ductal carcinoma of the breast: Relationship between beta-catenin, dishevelled and cyclin D1 expression.*Oncology* **2007**, *73*, 112–117.

71. Karayiannakis, A.J.; Nakopoulou, L.; Gakiopoulou, H.; Keramopoulos, A.; Davaris, P.S.; Pignatelli, M. Expression patterns of beta-catenin in *in situ* and invasive breast cancer. *Eur. J. Surg. Oncol.* **2001**, *27*, 31–36.

72. Currier, N.; Solomon, S.E.; Demicco, E.G.; Chang, D.L.; Farago, M.; Ying, H.; Dominguez, I.; Sonenshein, G.E.; Cardiff, R.D.; Xiao, Z.X.; *et al.* Oncogenic signaling pathways

activated in DMBA-induced mouse mammary tumors. *Toxicol. Pathol.* **2005**, *33*, 726–737.

73. Li, H.; Pamukcu, R.; Thompson, W.J. β-Catenin signaling: Therapeutic strategies in oncology.*Cancer Biol. Ther.* **2002**, *1*, 621–625.

74. Bhatia, D.; Thoppil, R.J.; Mandal, A.; Samtani, K.A.; Darvesh, A.S.; Bishayee, A. Pomegranate Bioactive Constituents Suppress Cell Proliferation and Induce Apoptosis in an Experimental Model of Hepatocellular Carcinoma: Role of Wnt/β-Catenin Signaling Pathway. *Evid. Based Complement. Altern. Med.* **2013**, *2013*, 371813.

75. Sharma, M.; Li, L.; Celver, J.; Killian, C.; Kovoor, A.; Seeram, N.P. Effects of fruit ellagitannin extracts, ellagic acid, and their colonic metabolite, urolithin A, on Wnt signaling. *J. Agric. Food Chem.* **2010**, *58*, 3965–3969.

76. Sadik, N.A.; Shaker, O.G. Inhibitory effect of a standardized pomegranate fruit extract on Wnt signalling in 1,2-dimethylhydrazine induced rat colon carcinogenesis. *Dig. Dis. Sci.* **2013**, *58*, 2507–2517.

77. Neuman, E.; Ladha, M.H.; Lin, N.; Upton, T.M.; Miller, S.J.; DiRenzo, J.; Pestell, R.G.; Hinds, P.W.; Dowdy, S.F.; Brown, M.; *et al.* Cyclin D1 stimulation of estrogen receptor transcriptional activity independent of cdk4. *Mol. Cell. Biol.* **1997**, *17*, 5338–5347.

78. Caldon, C.E.; Sutherland, R.L.; Musgrove, E. Cell cycle proteins in epithelial cell differentiation: Implications for breast cancer. *Cell Cycle* **2010**, *9*, 1918–1928.

79. Roy, P.G.; Thompson, A.M. Cyclin D1 and breast cancer. *Breast* **2006**, *15*, 718–727.

80. Kuramoto, T.; Morimura, K.; Yamashita, S.; Okochi, E.; Watanabe, N.; Ohta, T.; Ohki, M.; Fukushima, S.; Sugimura, T.; Ushijima, T. Etiology-specific gene expression profiles in rat mammary carcinomas. *Cancer Res.* **2002**, *62*, 3592–3597.

81. Shan, L.; He, M.; Yu, M.; Qiu, C.; Lee, N.H.; Liu, E.T.; Snyderwine, E.G. cDNA microarray profiling of rat mammary gland carcinomas induced by 2-amino-1-methyl-6-phenylimidazo[4,5-b]pyridine and 7,12-dimethylbenz[a]anthracene. *Carcinogenesis* **2002**, *23*, 1561–1568.

82. Papaconstantinou, A.D.; Shanmugam, I.; Shan, L.; Schroeder, I.S.; Qiu, C.; Yu, M.; Snyderwine, E.G. Gene expression profiling in the mammary gland of rats treated with 7,12-dimethylbenz[*a*]anthracene. *Int. J. Cancer* **2006**, *118*, 17–24.

83. Tetsu, O.; McCormick, F. Beta-catenin regulates expression of cyclin D1 in colon carcinoma cells.*Nature* **1999**, *398*, 422–426.

84. Seeram, N.P.; Adams, L.S.; Henning, S.M.; Niu, Y.; Zhang, Y.; Nair, M.G.; Heber, D. *In vitro*antiproliferative, apoptotic and antioxidant activities of punicalagin, ellagic acid and a total pomegranate tannin extract are enhanced in combination with other polyphenols as found in pomegranate juice. *J. Nutr. Biochem.* **2005**, *16*, 360–367.

85. Lansky, E.P.; Jiang, W.; Mo, H.; Bravo, L.; Froom, P.; Yu, W.; Harris, N.M.; Neeman, I.; Campbell, M.J. Possible synergistic prostate cancer suppression by anatomically discrete pomegranate fractions. *Investig. New Drugs* **2005**, *23*, 11–20.

86. De Kok, T.M.; van Breda, S.G.; Manson, M.M. Mechanisms of combined action of different chemopreventive dietary compounds: A review. *Eur. J. Nutr.* **2008**, *47*, 51–59.

87. Bode, A.M.; Dong, Z. Epigallocatechin 3-gallate and green tea catechins: United they work, divided they fail. *Cancer Prev. Res.* **2009**, *2*, 514–517.

88. Bishayee, A.; Bhatia, D.; Thoppil, R.J.; Darvesh, A.S.; Nevo, E.; Lansky, E.P. Pomegranate-mediated chemoprevention of experimental hepatocarcinogenesis involves Nrf2-regulated antioxidant mechanisms. *Carcinogenesis* **2011**, *32*, 888–896.

CHAPTER 9

Pomegranate Peel Extract Prevents Bone Loss in a Preclinical Model of Osteoporosis and Stimulates Osteoblastic Differentiation *in Vitro*

Mélanie Spilmont [1,2,3], Laurent Léotoing [1,2], Marie-Jeanne Davicco [1,2], Patrice Lebecque [1,2], Elisabeth Miot-Noirault [4], Paul Pilet [5], Laurent Rios [3], Yohann Wittrant [1,2] and Véronique Coxam [1,2,*]

[1]*Unité de Nutrition Humaine, CRNH Auvergne, UMR 1019, INRA, F-63000 Clermont-Ferrand, France*

[2]*Unité de Nutrition Humaine, Université d'Auvergne, Clermont Université, BP 10448, F-63000 Clermont-Ferrand, France*

[3]*GREENTECH SA Biopôle Clermont-Limagne, F-63360 Saint-Beauzire, France*

[4]*Imagerie Moléculaire et Thérapie Vectorisée, Université d'Auvergne, Clermont Université, UMR 990, INSERM, BP 10448, F-63000 Clermont-Ferrand, France*

[5]*Laboratoire d'Ingénierie Ostéo-Articulaire et Dentaire-LIOAD, Université de Nantes, UMR 791, INSERM, F-44042 Nantes, France*

ABSTRACT

The nutritional benefits of pomegranate have attracted great scientific interest. The pomegranate, including the pomegranate peel, has been used worldwide for many years as a fruit with medicinal activity, mostly antioxidant properties. Among chronic diseases, osteoporosis, which is associated with bone remodelling impairment leading to progressive bone loss, could eventually benefit from antioxidant compounds because of the involvement of oxidative stress in the pathogenesis of osteopenia. In this study, with *in vivo* and *ex vivo* experiments, we investigated whether the consumption of pomegranate peel extract (PGPE) could limit the process of osteopenia. We demonstrated that in ovariectomized (OVX) C57BL/6J mice, PGPE consumption was able to significantly prevent the decrease in bone mineral density (−31.9%; $p < 0.001$ *vs.* OVX mice) and bone microarchitecture impairment. Moreover, the exposure of RAW264.7 cells to serum harvested from mice that had been given a PGPE-enriched diet elicited reduced osteoclast differentiation and bone resorption, as shown by the inhibition of the major osteoclast markers. In addition, PGPE appeared to substantially stimulate osteoblastic MC3T3-E1 alkaline phosphatase (ALP) activity at day 7, mineralization at day 21 and the transcription level of osteogenic markers. PGPE may be effective in preventing the bone loss associated with ovariectomy in mice, and offers a promising alternative for the nutritional management of this disease.

Keywords: pomegranate peel extract; nutritional prevention; bone health; cell lines; animal model

1. INTRODUCTION

Bone is a dynamic organ that is continuously and precisely remodelled by the combined roles of osteoblasts and osteoclasts. Imbalance in bone remodelling is usually caused by a deregulated coupling between the main bone cells due to an increase in osteoclast resorption activity over the bone-formation rate by osteoblasts or by low bone turnover where both formation and resorption are reduced. This process leads to a common adult bone disease, *i.e.*, osteoporosis [1], which is a systemic skeletal disorder characterized by low bone mass and the structural deterioration of the microarchitecture, leading to bone fragility

[2]. Many attempts have been made to develop drugs for the prevention and treatment of this disease. Regarding nutritional prevention, despite the traditional focus on calcium and vitamin D, recent preclinical approaches have identified a multitude of nutrients endowed with bone sparing properties [3]. Indeed, it has been shown that polyphenolic compounds such as flavonoids, more precisely, genistein, daidzein, oleuropein, hydroxytyrosol, and others, exert an effect on osteoporosis [4,5,6,7,8,9], inhibit osteoclast differentiation [10,11,12] and stimulate osteoblast formation [13,14,15,16]*in vitro* and *in vivo*.

Among the main dietary sources of such micronutrients, pomegranate (*Punica granatum*, Punicaceae) has been used extensively in the folk medicine of many cultures [17]. While the relationship between pomegranate composition and its biological effects is not clearly elucidated, many recent advances have been made in the effort to discover the mechanisms involved and to identify the molecules responsible for these actions [18,19]. Various parts of the fruit have been shown to exert antioxidant [20,21], anti-inflammatory [22,23], anticarcigogenic [24,25,26], antiatherosclerotic [27,28], hypolipidemic [29,30], antidiabetic [31,32], antiviral, antibacterial, and antifungal activities [33,34,35], on cell lines, in preclinical models and in few human studies [36]. Indeed, the main properties of the pomegranate actually identified thus far are associated with its antioxidant capacity (three times higher than extracts of red wine and tea), and its composition including anthocyanins and tannins [37,38,39]. Actually, it has been reported that reactive oxygen species and low-grade inflammation, both of which contribute to the aging process [40,41,42,43,44,45], are involved in the aetiology of various degenerative diseases, such as osteoporosis [46,47,48,49,50].

The peel of the pomegranate represents almost 26%–30% of the fruit. This part of the fruit has the highest antioxidant capacity (92% of the total antioxidant activity of the fruit [51]) because of its large content of polyphenols such as punicalagin [52], flavonoids (anthocyanins, catechins and other complex flavonoids) and hydrolysable tannins (punicalin, pedunculagin, punicalagin, gallic and ellagic acid) [53]. Pomegranate peel tannins have been widely recognized and used traditionally for their medicinal properties, and several common ailments such as inflammation, diarrhoea, intestinal worms, cough, and infertility have been treated by exploiting pomegranate peel extract [17,54]. Consequently, the exceptional antioxidant and anti-inflammatory potential of pomegranate peel (together with a long history of use in folk medicine), associated with the need to develop new and innovative strategies for the management of osteoporosis, led us to investigate its role in the issue of bone health. In this study, using a pomegranate peel extract rich in tannins, we hypothesized that the consumption of pomegranate peel extract as a dietary supplement could exert a beneficial effect on bone biology. Moreover, using an *in vitro* cell culture model with osteoblasts and osteoclasts, we studied the cellular and molecular mechanisms of action that could be involved, and thanks to a murine model of osteoporosis we attempted to confirm observations at the cellular level.

2. MATERIALS AND METHODS

2.1. Pomegranate Peel Extract

Throughout the study, we used the pomegranate (*Punica granatum* L.) Wonderful cultivar, which is cultured in Israel and was purchased from POMONA (Clermont-Ferrand, France). It has sweet-tart taste, deep purple-red fruits with soft seeds and delicious vinous flavour. For all the *in vitro* and *in vivo*studies, we used the same batch of pomegranates that were stored at 4 °C. A careful sampling was performed for the analyses. Fresh fruits were peeled manually, and the peel was squeezed using a commercial turnmix blender (Philips HR2084, France) to obtain a homogenous puree. Pomegranate peel extract (PGPE) was obtained by hydro-alcoholic (ethanol/water* (mass/mass) 70/30 (*water from pomegranate peel)) extraction from peel puree, with a pomegranate/solvent ratio of 1/15 (w/w), at 90 °C, for 3 h. The extract was then filtrated through a 0.2 μm filter, the ethanol was evaporated and the extract was frozen at −20 °C, directly after preparation and until analysis and diet formulation. The administrated dose of 10 mg polyphenols/kg body weight/day on mice corresponds to a consumption of approximately 93 mg polyphenol/day in human (it is estimated that the spontaneous daily intake of polyphenols is approximately 1 g) [55]. Chemical characterization of PGPE composition was performed by AGROBIO (Rennes, France) and VEGEPOLYS INNOVATION (Angers, France) (see PGPE composition on Figure 1).

A

Pomegranate peel extract composition (g/100g) of dry matter	
Protein	1,54
Total sugar	81,62
Glucose	27,94
Fructose	32,79
Total fiber	-
Total phenols	15,06
Ellagic acid	2,89
Ellagic acid derivatives	5,08
Punicalagin	3,05
Gallic acid	0,89
Chlorogenic acid	0,021
Coumaric acid	0,047

B

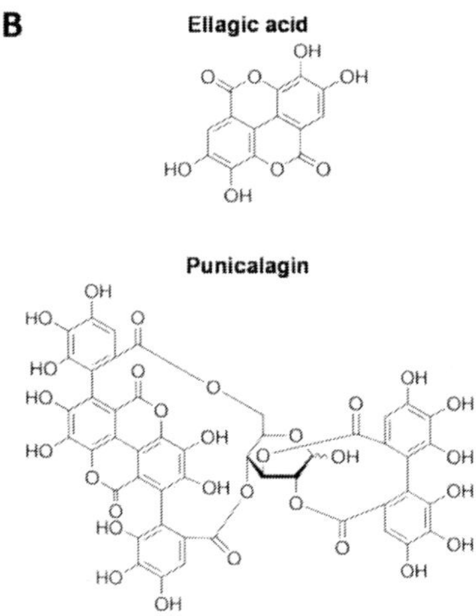

Figure 1. Composition of pomegranate peel extract (PGPE). (**A**) Chemical composition of PGPE (g/100 g dry matter). (**B**) Chemical structure of ellagic acid and punicalagin, the two major polyphenolic compounds of PGPE.

2.2. Animals Ethics

All animal procedures were approved by the institution's animal welfare committee (Comité d'Ethique en Matière d'Expérimentation Animale Auvergne: CEMEAA) and were conducted in accordance with the European guidelines for the care and use of laboratory animals (2010-63UE). The animals were housed in the animal facilities of the Human Nutrition Unit at INRA Research Centre [56], Agreement N°: C6334514).

2.3. In Vitro Study Design: Serum Production

Fifty 8-week-old female C57BL/6J mice were purchased from JANVIER (St Berthevin, France). After an acclimatization period of one week, the mice were randomly divided into 2 groups ($n = 25$ per group) denoted as the PGPE and control (CT) groups. Both groups were fed a control diet (AIN-93G). PGPE (50 mg polyphenol/kg body weight/day ellagic acid equivalent) was given to each animal in the PGPE group by forced oral administration, while the control group received the same amount of physiological saline, using the same procedure. After 10 days, venous blood from all the animals was collected under anaesthesia and centrifuged (3000 rpm for 5 min at room temperature). Serum supernatant was subsequently harvested. Then, serum from the same animal group was pooled, filtered at 45 μm, divided into aliquots and stored at −80 °C until use in the cell culture medium.

2.3.1. Cell Lines and Culture Conditions

Murine MC3T3-E1 osteoblast cells (ATCC, Washington, DC, USA) were seeded on collagen-1-coated plates (BD Biosciences, Bedford, MA, USA) at a density of 3×10^4 cells/cm^2. The cells were maintained in α-minimal essential medium (α-MEM; GIBCO, Paisley, UK) supplemented with 1% penicillin/streptomycin (GIBCO, Paisley, UK) and 10% heat-inactivated fetal bovine serum (FBS, Lonza, Levallois-Perret, France) (minimal medium). At 80% confluence, the cells were exposed to different conditions: minimal medium (C−) or minimal medium containing 50 μg/mL ascorbic acid and 10 mM β-glycerophosphate (C+; differentiation medium), in the presence of 10% heat-inactivated FBS (FBS) or 7.5% heat-inactivated FBS + 2.5% serum from mice that received physiological serum (Control) or pomegranate peel extract (PGPE), for 21 days.

In addition, RAW264.7, a murine monocytic cell line (ATCC, Washington, DC, USA), was used as an osteoclast model. The cells were seeded at a density of 1×10^4 cell/cm^2 and maintained in α-minimal essential medium (α-MEM; GIBCO, Paisley, UK) supplemented with 1% penicillin/streptomycin (GIBCO, Paisley, UK) and 10% heat-inactivated fetal bovine serum (FBS) (minimal medium). At 80% confluence, the cells were exposed to different conditions: minimal medium (C−) or minimal medium containing 50 ng/mL receptor activator of nuclear factor-kappa B ligand (RANK-L) (R&D Systems) (C+; differentiation medium), in the presence of 10% heat-inactivated FBS (FBS) or 7.5% heat-inactivated FBS + 2.5% serum from mice that received physiological serum (Control) or pomegranate peel extract (PGPE) for 4 days.

The replacement proportion of FBS by mice serum in these studies was determined by preliminary studies employing RAW264.7 and MC3T3-E1 cells to ensure that no modification of cell differentiation (cell viability, proliferation, specific activity, expression of major makers of cell type) was induced.

Both cell types were cultured at 37 °C in a humidified atmosphere of 5% CO_2 in air. The medium was replaced every 2 days. All experiments were repeated at least 3 times and reproduced 2 times.

2.3.2. Cell Proliferation

RAW264.7 and MC3T3-E1 cells were seeded in a 96-well plate at a density of 3.5×10^3 cells per well and then cultured for 48 h or 7 days with 2% serum (2% FBS or 1.25% FBS + 0.75% CT or PGPE mice serum). Cell viability was determined by an XTT-based method using Cell Proliferation Kit II (Sigma-Aldrich, St. Louis, MO, USA) according to the manufacturer's recommendations. The OD was determined at 450 nm.

2.3.3. Alkaline Phosphatase (ALP) Activity Measurement

The enzymatic activity of ALP was measured in osteoblasts at days 0, 2, 7 and 14, according to previously published methods [57], adapted to our experimental conditions. Briefly, the osteoblast cultures were rinsed twice with PBS (Sigma-Aldrich, 38297 Saint-Quentin Fallavier, France) prior to freezing at −20 °C. The cells were then lysed by freeze–thaw cycles and homogenized in diethanolamine/magnesium chloride hexahydrate buffer (pH 9.8; Sigma-Aldrich). The cell lysates (10 μL) were added to 200 μL of p-nitrophenyl phosphate solution (6 mg/mL) (Sigma-Aldrich, St. Louis, MO, USA). Absorbance was assessed at 405 nm, 30 °C, every 150 s for 30 min, using an ELX808 microplate reader (BioTek Instruments Inc, Winooski, VT, USA). Protein measurement was performed using a BioRad protein assay (BioRad, Munich, Germany). ALP activity was expressed as micromoles of*p*-nitrophenol per hour per milligram of protein [58].

2.3.4. Tartrate-Resistant Acid Phosphatase (TRAP) Activity Measurement

TRAP activity was assayed according to the standard method using *p*-nitrophenyl phosphate as a substrate [59]. The medium was removed from osteoclasts cultured in 24-well plates after 3 days, and cell lysates were prepared using NP40 lysis buffer. The samples were incubated in assay buffer (125 mM sodium acetate buffer (pH 5.2), 100 mM *p*-nitrophenyl phosphate (Sigma-Aldrich, St. Louis, MO, USA), and 1 mM L(+) sodium tartrate). The production of *p*-nitrophenol was determined at 405 nm, 37 °C and expressed as the mean OD per minute per milligram protein.

2.3.5. Mineralization

Osteoblasts seeded in 12-well plates were cultured for 21 days in minimal medium containing 50 μg/mL ascorbic acid and 10 mM β-glycerophosphate to investigate bone nodule formation. Extracellular matrix calcium deposits were stained using Alizarin red dye, as previously described [58]. Osteoblasts were fixed with ice-cold 70% (*v*/*v*) ethanol and then stained with Alizarin Red S (40 mM) (Sigma-Aldrich, St. Louis, MO, USA) for 30 min. Mineralization was evaluated by the light microscopy and image processing software ImageJ (National Institutes of Health, Bethesda, MD, USA).

2.4. In Vivo Study Design

Thirty 8-week-old female C57BL/6J mice were purchased from JANVIER (St Berthevin, France). After an acclimatization period of one week, the mice were randomly divided into 3 groups ($n = 10$ per group). Two groups were surgically ovariectomized (OVX), under ketamine/xylazine anaesthesia, while the other animals were sham operated (SH). Paracetamol was added to the water for 24 h post-surgery for pain limitation [60,61].

Control mice (SH Control and OVX Control) were fed a standard AIN-93G diet, while the last group (OVX PGPE) received the standard diet enriched with 2 g/kg of PGPE (*i.e.*, 10 mg polyphenols (ellagic acid equivalents)/kg body weight/day). The diets were purchased from SAFE (Scientific Animal Food and Engineering, Augy, France).

The animals were housed in a controlled environment (12:12 h light-dark cycle, at 20–22 °C, with 50%–60% relative humidity, 1 mouse per plastic cage), with free access to water. Body weight was measured every two days during the study period. Body composition was assessed at the beginning and at the end of the study, using a QMR (Quantitative magnetic resonance) EchoMRI-900™ system (EchoMRI LLC, Houston, TX, USA), without any anaesthesia or sedation. After the 30-day treatment, the mice were sacrificed. The liver, spleen and uterus were weighted. Femurs and tibias were harvested and stored at −80 °C prior to investigation.

2.4.1. Bone Mineral Density (BMD) Analysis

Bone morphological analysis was performed using an eXplore CT 120 scanner (GE Healthcare, Little Chalfont, United Kingdom). After the removal of soft tissues, the left femurs were placed in PBS buffer with 10% formaldehyde at 4 °C for one week and scanned. Acquisition consisted of 360 views acquired in 1 soft tissues, the left femurs were placed in PBS buffer with 1 millisecond exposure/view and X-ray tube settings of 100 kV and 50 mA. CT images were reconstructed using a modified conebeam algorithm with an isotropic voxel of 0.045 x 0.045 x 0.045 mm^3. CT scans were analysed using MicroViewH version 2.3 software (General Electric Healthcare Bio-Sciences, Pittsburgh, PA, USA). A hydroxyapatite calibration phantom (SB3, Gamex RMI, WI, USA) was used to convert grey-scale levels to hydroxyapatite density values. The trabecular bone of the distal femur was selected for bone mineral density and bone volume fraction (BVF = Bone Volume/Total volume) analyses by fitting a cylindrical region (r = 0.7 mm) of interest in the center of the femur, starting 0.1 mm proximally from the growth plate and extending a further 0.32 mm in the proximal direction. Bone mineral density was estimated as the mean converted grey-scale level within the region of interest.

2.4.2. Bone Micro-Architecture Analysis

To perform a measurement, the specimen was mounted on a turntable that could be shifted automatically in the axial direction (angular step: 0.675 /reconstruction angular range: 186.30). An aluminium filter (0.5 mm thick) was placed between the X-ray source and the sample. The micro-architecture (secondary spongiosa) was analysed using X-ray radiation provided by a micro-CT SkyScan 1072 (BRUKERMICROCT, Kontich, Belgium). Pictures of 1024 × 1024 pixels were obtained using 37 kV and 215 mA. According to the camera settings, the final pixels measured 5.664 mm, leading to a voxel of 1.817 × 1027 mm^3. The calculation of histomorphometric parameters was performed using the CTAn H version 1.11 (BRUKER MICROCT, Kontich, Belgium), and

NreconH software version 1.6.1.7 [62] (BRUKER MICROCT, Kontich, Belgium).

2.5. Taqman Low Density Arrays (TLDA)

For each experimental group, four sets of extractions were performed with two tibias pooled in each. Frozen bones were ground in liquid nitrogen to obtain a fine powder. Then, total RNA was extracted according to the manufacturer's instructions from either bone powder for the *in vivo* experiment or cell culture, using Trizol reagent (Invitrogen, Life Technologies, Saint Aubin, France). RNA was converted to cDNA using the high capacity cDNA reverse transcription (RT) kit (Applied Biosystems). The resulting cDNA was used for TaqMan® low-density arrays (TLDAs) (Applied Biosystems 7900HT real-time PCR system). Relative expression values were calculated using the comparative threshold cycle ($2^{-\Delta\Delta}$CT) according to the Data Assist software (Applied Biosystems, Life Technologies, Saint Aubin, France). 18S, GAPDH and actin served as housekeeping genes.

2.6. Statistical Analysis

Results are expressed in the form of mean with standard error (SEM). All the data were analysed using XLSTAT Pro software [63] (Addinsoft, 40 rue Damrémont, Paris, France). *In vitro* data were analysed using a two-way analysis of variance (ANOVA) to test for differences among groups. If the result was found to be significant ($p < 0.05$), the fisher multiple comparison test was then performed to determine the specific difference between means. TLDA data and *in vivo* data were analysed using non-parametric Kruskal–Wallis one-way analysis of variance, which allows testing for differences among groups. If the result was found to be significant ($p < 0.05$), the Mann-Whitney U test was performed to determine the specific difference between means.

3. RESULTS

To determine the accurate potential effect of pomegranate peel extract (PGPE) on bone metabolism, we first developed an *in vitro* model relevant to nutritional conditions. This model allowed us to consider plasma modifications elicited by PGPE consumption (including both production of metabolites from the pomegranate polyphenols and any systemic effects of those molecules) and to study the impact of pomegranate on bone cell cultures *in vitro*. For that purpose, serum from mice that had been given PGPE or physiological saline (control) was introduced into the medium of bone cell cultures (in which 25% of FBS was replaced by mouse serum: 7.5% heat-inactivated FBS + 2.5% serum from mice).

3.1. nPGPE Promoted Osteoblast Differentiation

3.1.1. Cell Viability

To insure that mouse sera (harvested from mice fed either a standard diet (Control) or the control diet enriched with PGPE) were not deleterious for MC3T3-E1 cells, a cell viability test was performed. Incubation with serum from PGPE animals did not modify the proliferation of the MC3T3-E1 cells, on days 2 and 7, in comparison with cells cultured with serum from CT mice (Figure 2A).

3.1.2. Effect of PGPE on ALP Activity and Mineralization of MC3T3-E1 Cells

The kinetic measurement of ALP activity in cells treated with mice serum raised in either Control or PGPE conditions on days 2, 7 and 14 was first assessed to investigate the effect of the tested diets on osteoblast differentiation. With regard to the Control condition (FBS), as expected, the osteoblasts treated with optimized medium (C+) exhibited higher ALP activity compared to osteoblasts exposed to a minimal medium (C−), as of day 2 (+72%), day 7 (+374%; $p < 0.01$) and day 14 (+139%) (data not shown). The cells that had been exposed to PGPE mice serum exhibited higher values compared to the Control condition (+46% at day 7 and +16% at day 14; $p < 0.05$). (Figure 2B).

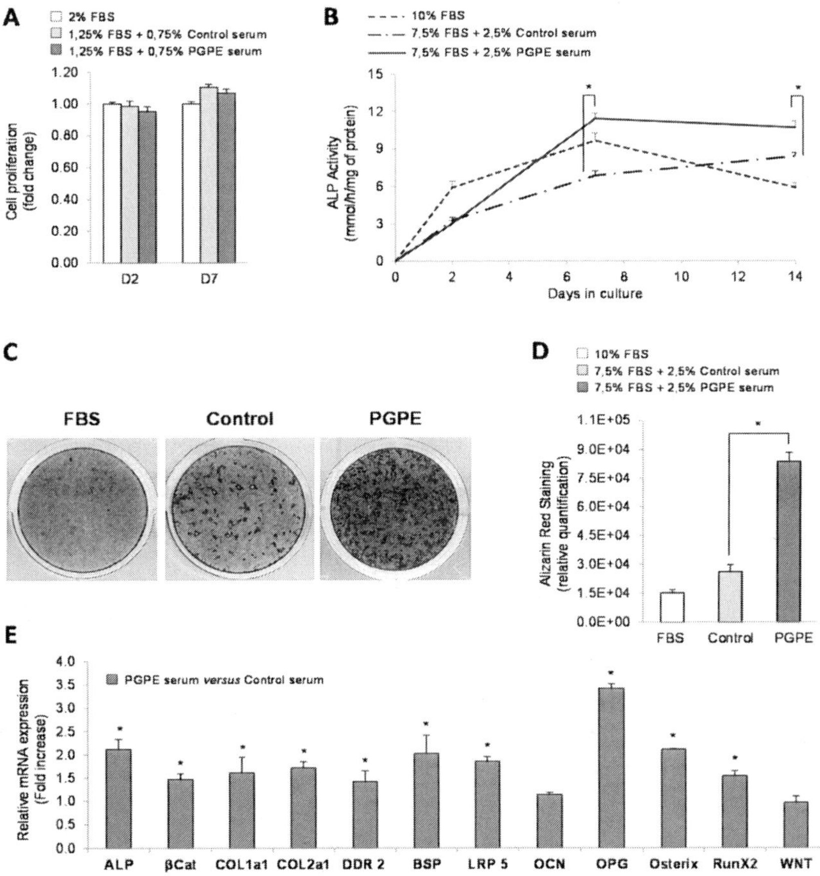

Figure 2. Osteoblast viability, differentiation and transcription in culture, after exposure to serum harvested from mice raised either under control conditions or treated with pomegranate peel extract. *$p < 0.05$ PGPE *versus* Control condition. (**A**) Cell proliferation test performed on MC3T3-E1 cells cultured in minimal medium (2% Fetal Bovine Serum, FBS), supplemented with serum from mice given either physiological serum (1.25% FBS + 0.75 Control mice serum) or pomegranate peel extract, PGPE (1.25% FBS + 0.75 PGPE mice serum), after 2 and 7 days of treatment. Results are expressed as mean ± standard error of the

mean (SEM) (*n* = 6). (**B**) Alkaline phosphatase, ALP activity of MC3T3-E1 cells cultured in optimized medium with 10% FBS or with 7.5% FBS + 2.5% serum from mice fed either physiological serum (Control) or pomegranate peel extract (PGPE), after 0, 2, 7 and 14 days of treatment. Results are expressed as mean ± SEM (*n* = 8). (**C**,**D**) Mineralized nodules stained with alizarin red S on Day 21 in MC3T3-E1 cells cultured in optimized (C+) medium with 10% FBS or with 7.5% FBS + 2.5% serum from mice exposed to either physiological serum (Control) or pomegranate peel extract (PGPE). (**C**) Representative image of alizarin red staining for each condition. (**D**) Relative quantification of alizarin red staining with picture analyser software ImageJ. Results are presented as mean ± standard deviation, SD (*n* = 6). (**E**) Transcriptomic analysis of the main osteoblast markers in MC3T3-E1 cells cultured in optimized medium after 7 days of treatment with serum from mice exposed to either physiological serum (Control) or pomegranate peel extract (PGPE). Transcriptomic analysis of MC3T3-E1 mRNA levels determined by Taqman Low-Density Arrays are presented as fold change compared to the Control condition (fold change = 1) as mean ± SD (*n* = 6). Genes: ALP: alkaline phosphatase; βCat: β-catenin ; Coll1a1: collagen 1a1; Coll2a1: collagen 2a1; DDR2: discoidin domain receptor 2; BSP: bone sialoprotein; Lrp5; OCN: osteocalcin; OPG: osteoprotegerin; osteri RunX2 and Wnt.

In addition to this ALP activity and to target the effect of PGPE on bone mineralization, we analysed calcium deposition by MC3T3-E1 cells under differentiated conditions (C+) and when treated with either PGPE mouse serum or Control mouse serum for 21 days. As shown in Figure 2C, and as expected, the osteoblasts treated with PGPE mouse serum were able to produce many more calcium nodules after 21 days of culture than osteoblasts treated with Control mouse serum (Figure 2D).

3.1.3. Effect of PGPE on transcriptional activity of MC3T3-E1 cells

To better understand the effect of PGPE on osteoblasts, transcriptional analyses of major differentiation and transcription factors were performed. Due to the large number of targets, only the most relevant and significant ones are presented. The expression level of mRNA extracted from MC3T3-E1 cells after 7 days of culture with serum from PGPE mice was compared to the expression level of mRNA from cells treated with serum from Control mice at the same culture time (Fold increase value = 1). As shown in Figure 2E, PGPE led to a significant ($p < 0.05$) increase in the expression of the major osteogenic markers such as ALP (2.12-fold), collagen 1a1 and 2a1 (Coll1a1 and Coll2a1) (1.62-fold and 1.72-fold respectively), bone sialoprotein (BSP) (2.02-fold), osteocalcin (1.14-fold; not significant) and osteoprotegerin (OPG) (3.41-fold). In addition, discoidin domain receptor 2 (DDR2), a cell surface receptor that plays critical roles in the collagen control of osteoblast differentiation, was up-regulated by PGPE (1.42-fold; $p < 0.05$). Two members of the Wnt/β-catenin signalling pathway that regulate osteoblast differentiation and bone formation were also up-regulated by PGPE: β-catenin (1.47-fold; $p < 0.05$) and LRP5 (1.85-fold; $p < 0.05$), but not Wnt (0.97 fold). Finally, PGPE stimulated the expression of the two master osteoblastic transcription factors: osterix and runX2 (2.11- and 1.54-fold respectively; $p < 0.05$).

3.2. PGPE Did Not Affect Osteoclast Activity but Impaired Transcription

3.2.1. Cell Viability

With regard to the osteoclast model (RAW264.7 cells), proliferation was significantly increased at day 2 in all conditions, in comparison to day 1. However, mouse serum induced a slight decrease in osteoclast proliferation on the first culture day compared to what was observed in the FBS condition. No significant difference was noticed between the PGPE and Control serum conditions (−36% and −21% respectively; $p < 0.5$) (Figure

3A). Then, on day 2, we observed that in the PGPE condition, osteoclast proliferation was reduced compared to the Control mouse serum condition (−26%, $p < 0.01$).

3.2.2. Effect of PGPE on TRAP Activity of RAW264.7 Cells

We assessed the TRAP activity in RAW264.7 cells after three days of culture to determine whether serum from PGPE mice was able to reduce osteoclast differentiation. As described in Figure 3B, exposure to mouse serum was associated with an increase of such a parameter. However, PGPE did not elicit any change compared to observations using serum from control mice.

3.2.3. Effect of PGPE on Transcriptional Activity of RAW264.7 Cells

To further investigate the effect of PGPE on bone resorption, the expression levels of specific differentiation markers and transcription factors were assessed. A TLDA analysis was therefore performed on mRNA extracted from RAW264.7 cells treated with mouse serum (Control or PGPE) under the differentiated condition (C+) after three days of culture. The expression of the most relevant targets is shown in Figure 3C. The exposure to serum from PGPE mice was compared to the results of Control treated cells (Fold increase value = 1). The TLDA data indicate that the exposure of RAW264.7 differentiated cells to serum from PGPE mice was associated with a slight, but significant ($p < 0.05$), decrease in the mRNA levels of many crucial osteoclast-specific markers, such as calmodulin (CaM) (0.84-fold), chemokine receptor 2 (CCR2) (0.80-fold), cathepsin K (CTSK) (0.78-fold) and metalloproteinase 9 (MMP9) (0.88-fold) (Figure 3C). Consistently with this effect, down-regulated expression of a downstream NF-κB subunit, RelB, was observed in the PGPE condition (0.77-fold; $p < 0.05$). Moreover, the expression of interferon β1 (INFβ1), an osteoclastogenesis inhibitor, was induced (1.58-fold; $p < 0.05$) in the PGPE condition.

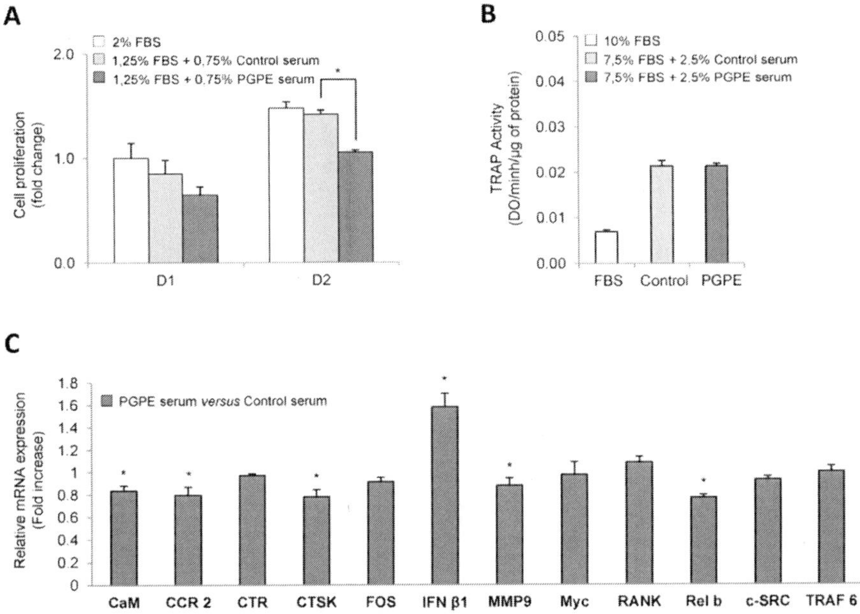

Figure 3. Osteoclast viability, differentiation and transcription in culture after exposure to serum harvested from mice either raised under control conditions or treated with pomegranate peel extract. *$p < 0.05$ PGPE *versus* Control condition. (**A**) Cell proliferation test performed on RAW264.7 cell line cultured in minimal medium (2% Fetal Bovine Serum, FBS), supplemented with serum from mice exposed to either physiological serum (1.25% FBS + 0.75% Control mice serum) or pomegranate peel extract, PGPE (1.25% FBS + 0.75 PGPE mice serum) after 1 and 2 days of treatment. Results are expressed as mean ± standard error of the mean (SEM) (n = 6). (**B**) Tartrate resistant acid phosphatase, TRAP activity of RAW264.7 cells cultured in optimized medium with 10% FBS or with 7.5% FBS + 2.5% mice serum from animals fed either physiological serum (Control) or pomegranate peel extract (PGPE), after 3 days of treatment. Results are expressed as mean ± SEM (n = 8). (**C**) Transcriptomic analysis of the main osteoclast markers in RAW264.7 cells cultured in optimized medium after 2 days of treatment with serum from mice exposed to either physiological serum (Control) or pomegranate peel extract (PGPE). Transcriptomic analysis of the Raw264.7 mRNA levels determined by Taqman Low density Arrays are presented as fold change compared to Control condition (fold change = 1) as mean ± standard deviation (SD) (n = 6). Gene: CaM: calmodulin; CCR2: chemokine receptor 2; CTR: calcitonin receptor; CTSK: cathepsin K; FOS; IFN β1: interferon β1; MMP9: metalloproteinase 9; Myc, RANK: receptor activator of nuclear factor-κB; Relb; c-SRC: and TRAF6.

3.3. Confirmation of Cell Culture Results Using a Physiological in Vivo Approach

These *ex vivo* results were completed by an *in vivo* investigation of the possible effect of pomegranate using a well-described model of osteoporosis induction by ovariectomy on mice.

3.3.1. Validation of the Animal Model

During this study, the mean body and uterine weights, as well as the lean and fat mass, were measured to ensure that our model conformed to published data. In Table 1, we observed that anthropometric parameters (total weight, fat and lean mass) increased in all the experimental groups throughout the experiment. No significant difference was observed between groups, SH Control *vs.*OVX Control or OVX PGPE *vs.* OVX Control. As expected, ovariectomy significantly induced marked uterine atrophy in the OVX Control group (Table 1). This effect on the uterus was not prevented by pomegranate peel extract consumption.

3.3.2. PGPE Consumption Improved Bone Mineral Density in Ovariectomized Mice

Bone morphological parameters using micro-CT analysis were first assessed to evaluate the respective effects of ovariectomy and PGPE consumption on mice. We confirmed the acknowledged effect of oestrogen deficiency on femoral BMD by comparing the OVX Control animals and the SH Control group. Indeed, the OVX Control mice had a significant lower BMD ($p < 0.001$) than the SH Control mice (−35.6%) (Figure 4A). PGPE consumption was associated with the prevention of such a bone decrease in OVX animals (+31.9%; $p < 0.001$ compared to the OVX Control animals).

Table 1. Effect of ovariectomy and pomegranate peel extract consumption for 30 days on physiological parameters in mice.

	Sham	Ovariectomized	
	Normal Diet	Normal Diet	PGPE Diet
	(SH)	(OVX)	(OVX+PGPE)
Initial body weight (g)	18.9 ± 0.6	19.4 ± 0.6	19.0 ± 0.5
Initial lean mass (% body weight)	87.6 ± 3.8	87.1 ± 1.2	87.1 ± 0.9
Initial fat mass (% body weight)	6.6 ± 2.1	7.9 ± 1.3	8.0 ± 1.2
Final body weight (g)	19.3 ± 0.8	20.3 ± 1.2	19.9 ± 0.9
Final lean mass (% body weight)	88.9 ± 7.3	92.3 ± 3.0	91.6 ± 2.7
Final fat mass (% body weight)	9.8 ± 2.2	9.3 ± 1.6	9.0 ± 2.0
Uterine weight (mg)	84.4 ± 3.2	23.1 ± 2.8 #	23.7 ± 2.8

Mice were sham-operated, SH or ovariectomized, OVX and fed standard diet AIN-93 (Control) or standard diet enriched with 2g/kg of pomegranate peel extract, PGPE representing a dose of 10 mg polyphenols/kg body weight/day (ellagic acid equivalent) for 30 days. Each group contained 10 mice. Values are presented as means ± standard error of the mean, SEM. # $p < 0.005$ OVX Control versus SH Control group.

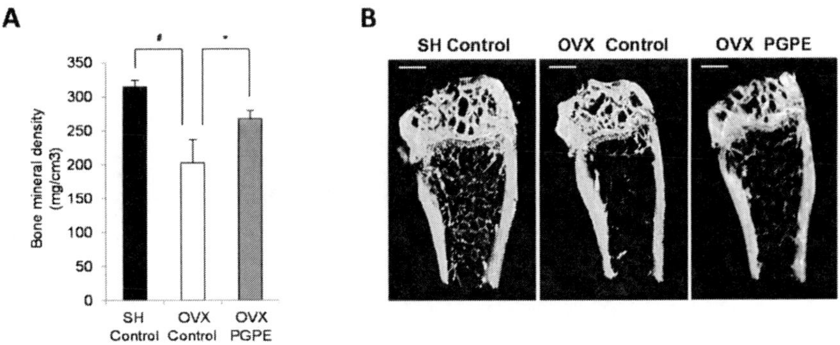

Figure 4. Effect of ovariectomy and pomegranate peel extract consumption for 30 days on bone parameters in mice. (**A**) Bone mineral density analysis of left femur by X-ray radiation micro-CT. Mice were sham-operated, SH or ovariectomized, OVX and fed standard diet AIN-93 (Control) or standard diet enriched with 2g/kg of pomegranate peel extract, PGPE representing a dose of 10 mg polyphenols/kg body weight/day (ellagic acid equivalent) for 30 days. Each group contained 10 mice. Results are presented as means ± standard error of the mean, SEM. # $p < 0.005$ OVX Control versus SH Control group. * $p < 0.05$ OVX PGPE *versus* OVX Control. (**B**) Representative microCT images of the distal left femur for each group: SH Control, OVX Control and OVX PGPE. Scale bars, 1 mm.

3.3.3. PGPE Preserved Bone Microarchitecture in Ovariectomized Mice

We also observed a significant effect on the trabecular bone microarchitecture of the distal femur, as shown in Figure 4B. Again, as expected, bone microarchitecture was significantly impaired by ovariectomy, with an increase in trabecular separation (TbSP +26.1%; $p < 0.001$) as well as total porosity (Po tot +3.5%; $p < 0.01$) in comparison with the SH Control group (Table 2). Significant decreases in trabecular number (TbN: −25.0%; $p < 0.05$), bone per cent volume (BV/TV: −23.6%; $p < 0.01$), connectivity density (Conn Dn: −19.7%; $p < 0.01$) and bone surface density (BS/TV: −23.9%; $p < 0.05$) were also observed. In addition, the consumption of PGPE exhibited a protective effect on bone microarchitecture, as PGPE consumption improved significantly ($p < 0.01$): BV/TV (+23.1%), TbN (+22.6%), structure model index (SMI: −7.3%), Po tot (−2.52%) and BS/TV (+17.1%) compared to the values measured in the OVX Control animals.

Table 2. Effect of ovariectomy and pomegranate peel extract consumption for 30 days on bone micro-architectural parameters in mice.

	Sham	Ovariectomized	
	Normal Diet	Normal Diet	PGPE Diet
	(SH)	(OVX)	(OVX+PGPE)
BV/TV (%)	12.7 ± 0.9	9.7 ± 0.6 #	11.9 ± 0.76 *
TbTh (mm)	0.1 ± 0.0	0.1 ± 0.0	0.1 ± 0.0
TbN (mm^{-1})	2.0 ± 0.1	1.5 ± 0.1 #	1.8 ± 0.07 *
TbSP (mm)	0.2 ± 0.0	0.3 ± 9.97E-03 #	0.3 ± 0.0
SMI	2.5 ± 0.1	2.6 ± 0.1	2.4 ± 0.04 *
Conn Dn (mm^{-3})	278.4 ± 26.5	223.5 ± 8.8	224.3 ± 12.7
Tb Pf (mm^{-1})	25.5 ± 1.1	27.7 ± 1.0	23.2 ± 1.2
Po tot (%)	87.3 ± 1.0	90.3 ± 0.61 #	88.1 ± 0.76 *
BS/TV (mm^{-1})	8.9 ± 0.5	6.7 ± 0.32 #	7.9 ± 0.31 *

Mice were sham-operated, SH or ovariectomized, OVX and fed standard diet AIN-93 (Control) or standard diet enriched with 2g/kg of pomegranate peel extract, PGPE representing a dose of 10 mg polyphenols/kg body weight/day (ellagic acid equivalent) for 30 days. Each group contained 10 mice. Values represent means ± standard error of the mean, SEM. # $p < 0.005$ OVX Control *versus* SH Control group. * $p < 0.05$ OVX PGPE *versus* OVX Control. Abbreviations: BV/TV: Per cent bone volume; TbTh: trabecular thickness; TbN: trabecular number; TbSP: trabecular spacing; SMI: structure model index; ConnDn: connectivity density; Tb Pf: trabecular pattern factor; Po tot: total porosity; BS/TV: bone surface density.

3.3.4. Pomegranate Intake Was Associated with an Improved Expression Profile of Specific Bone Markers

To elucidate the possible molecular mechanisms involved in this effect on bone morphology and to confirm the *in vitro* findings, we performed a transcriptomic analysis on bone tissue samples using Taqman Low Density Arrays (TLDA, Applied Biosystems, Life Technologies, Saint Aubin, France). Because of the large number of genes modulated, only the most important bone markers are presented. Data on the OVX PGPE group were compared to the data measured for the OVX Control group. The TLDA data on osteoclasts are shown in Figure 5A. Accordingly, PGPE consumption significantly ($p < 0.05$) down-regulated the mRNA levels of some of the major osteoclast-specific markers: define (0.39 fold) and integrin β3 (ITG b3: 0.56 fold). Fos, a crucial transcription factor for osteoclast differentiation, was also inhibited in the PGPE group (0.22 fold; $p < 0.05$). Finally, we observed a trend toward down-regulated expression in the OVX PGPE group of a key enzyme for osteoclast activity involved in collagen degradation: the metallo-proteinase 2 (MMP2) (PGt: 0.55; not significant). In addition, regarding the expression of osteoblast-specific markers, the TLDA data presented in Figure 5A show that PGPE consumption by ovariectomized mice was able to significantly improve the expression of a co-receptor implied in Wnt/β-catenin signalling in the osteoblast, LRP5 (1.62 fold; $p < 0.05$), in comparison with the OVX Control group. However, the other osteoblast markers investigated were not significantly modulated by dietary PGPE.

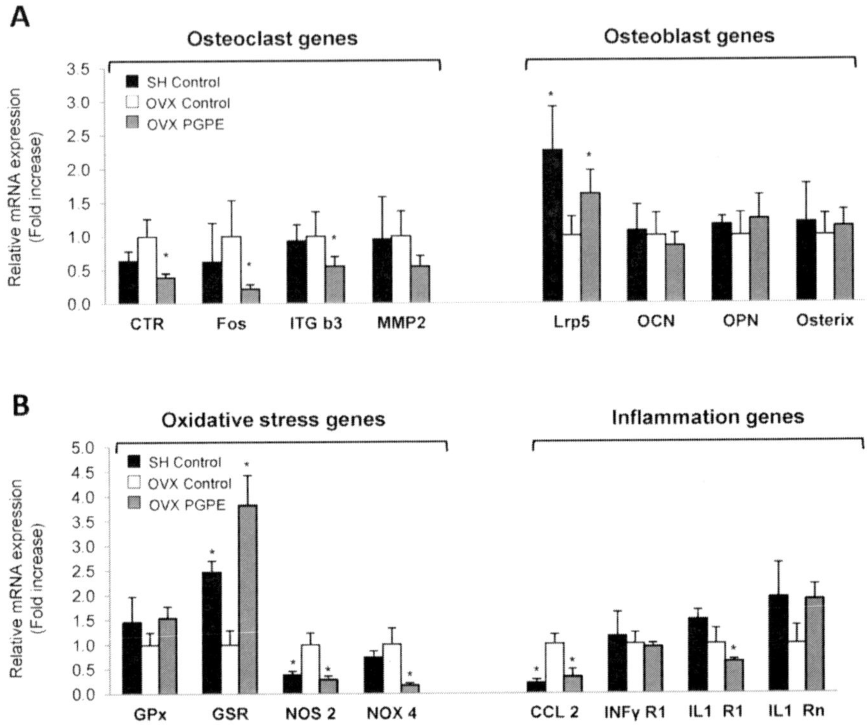

Figure 5. Expression profile analysis of bone (**A**), oxidative stress and inflammation (**B**) markers in femoral bone from mice fed the control diet (Sham-operated, SH Control. Ovariectomized, OVX Control) or exposed to pomegranate peel extract (OVX PGPE, diet containing 2 g/kg of PGPE), for 30 days. Transcriptomic analysis of bone tissue mRNA levels determined by Taqman Low density Arrays are presented as fold change compared to OVX Control group (fold change = 1). Results are expressed as mean ± standard deviation, SD (n = 8). * $p < 0.05$. (**A**) Osteoclast genes: CTR: calcitonin receptor; Fos; ITG b3: integrin β3; MMP2: metalloproteinase 2. Osteoblast genes: Lrp5; OCN: osteocalcin; OPN: osteopontin; RunX2. (**B**) Oxidative stress genes: GPx: glutathione peroxidase; GSR: glutathione reductase; NOS2: nitric oxide synthase 2; NOX4: NADPH oxidase 4. Inflammation genes: CCL2: Chemokine (C-C motif) ligand 2; INFγR1: interferon gamma receptor 1; IL1-R1: interleukin 1 receptor 1; IL1-Rn: interleukin 1 receptor antagonist.

3.3.5. PGPE Enhanced Bone Inflammatory and Oxidative Status Marker Expression in Ovariectomized Mice

To obtain further conclusions regarding the potent bone-sparing effect of pomegranate consumption, we investigated the expression of major inflammation and oxidation markers, which are known to exacerbate bone resorption. We therefore analysed the transcription of some inflammatory and oxidation markers in bone tissue samples using Taqman Low Density Arrays (TLDA, Applied Biosystems, Life Technologies, Saint Aubin, France). Regarding the expression of oxidation markers in the bone microenvironment, dietary PGPE enhanced bone oxidative stress in ovariectomized mice compared to mice given the standard diet, as shown in Figure 5B. PGPE consumption up-regulated the expressions of antioxidant enzymes (GPx (1.54-fold; trend) and GSR (3.82-fold; $p < 0.05$)), while the expression of the ROS synthesis enzyme NOS2 was down-regulated (0.16-fold; $p < 0.001$). In addition, PGPE consumption by ovariectomized mice was able to prevent the instigation of a low-grade inflammatory status compared to OVX Control animals, through the down-regulated expression of CCL2 (0.33-fold; $p < 0.05$) and IL1-R1 (0.64-fold; $p < 0.001$) and the up-regulated expression of IL1-Rn (1.89-fold; $p < 0.001$), again supporting our previous results.

4. DISCUSSION

Taken together, these results outline the potential effect of pomegranate peel extract consumption on bone health. Indeed, using an *ex vivo* investigation that makes it possible to address the overall physiological conditions (by considering the specific metabolism of pomegranate peel extract polyphenols and the possible systemic effects of the consumption of such ingredients) coupled with a well-described preclinical model, we demonstrated that dietary PGPE could improve bone mass and microarchitecture with transcriptional changes in bone tissue during osteoporosis establishment.

Our PGPE is composed of free glucose and fructose (~80%) and polyphenols (~15%) comprising mostly ellagitannins: ellagic acid (~2.9%), derivatives (~5.1%) and punicalagin (~3.1%). In the body, the native ellagitannin forms of PGPE are converted to ellagic acid through a (non-enzymatic) reaction due to acid hydrolysis, which occurs in the caecum at a pH of 8. Punicalagin catabolism leads to the release of ellagic acid, which is largely absorbed and converted by the bacterial microflora in the intestinal lumen to produce the main metabolites of punicalagin, urolithins (mainly forms A and B in humans). While ellagitannins are not found in either the plasma or the urine after the ingestion of pomegranate, their metabolites (ellagic acid and urolithins) are present [23,64,65]. In our studies, the maximal doses of PGPE administered were 10 (*in vivo* study) or 50 (*in vitro* study) mg ellagic acid equivalents/kg body weight/day. These doses were safe, as it has been reported that the oral administration of 600 mg/kg/day of pomegranate extract to rats, corresponding to a dose of 180 mg punicalagin/kg/day for 90 days, was devoid of any observed adverse effect level (NOAEL) [66]. Thus, using serum from mice fed with PGPE or physiological serum (Control) on *in vitro* bone cell culture, we attempted to reproduce the bone nutritive microenvironment induced by PGPE nutritional supplementation in the animals. Indeed, in physiological conditions, bone cells are never in direct contact with the native form of those molecules.

Using our *ex vivo* MC3T3-E1 cell line model, we found that PGPE significantly stimulated osteoblastogenesis, as demonstrated by increased ALP activity and calcium nodule formation. The transcriptional analysis of major osteoblast lineage markers was consistent with those two markers of osteoblast activity: ALP, BSP, Coll1a1, Coll2a1, DDR2, OCN, OPG were all significantly up-regulated by PGPE. Similarly, an increase in the expression of the two main osteoblastic transcription factors, osterix and runx2, allowed us to hypothesize that transcriptional changes might contribute to the improved osteoblast function [67,68]. Previous works conducted using MC3T3-E1 cells have suggested an osteogenic role of a pomegranate ethanolic extract [20] or pomegranate concentrate powder extract [69] underlined by increased ALP activity, even though the authors did not consider the particular metabolism of pomegranate micronutrients.

We also investigated, for the first time, the effects of PGPE on bone cell resorption based on an *in vitro*concept using RAW264.7. We found that serum from PGPE mice was able to inhibit RANKL-induced osteoclast differentiation, as shown by the down-regulation of the expression of specific osteoclast markers (calmodulin, CCR2, calcitonin receptor, cathepsin K, MMP-9) [70]. We observed a consistent decreased expression of relB, the downstream NF-κB subunit responsible for osteoclast differentiation [71]. In addition, the expression of interferon β1, an inhibitor of osteoclastogenesis [72], was up-regulated, thus providing a potential explanation for the PGPE-related inhibition of osteoclastogenesis. Consistently with our data, an *in vitro* anti-resorptive activity of pomegranate pericarp extract has also been suggested, as it was shown to inhibit the receptor activator of nuclear factor-κB ligand (RANKL) expression in MG-63 human osteosarcoma cells [73]. To date, except for the work conducted with ellagitannin (furosin, extracted from Euphorbia helioscopia) on osteoclasts [74], showing the suppression of RANKL-induced osteoclast differentiation and function through the inhibition of MAP kinase activation, no data have targeted the role of pomegranate polyphenols on RAW264.7 osteoclasts.

In agreement with those results, *in vivo* investigation here also demonstrated that PGPE consumption was associated with the decrease of pro-resorption marker expression using Taqman Low density Arrays.

In addition to the direct effects on bone demonstrated by our *in vitro* data, PGPE could modulate bone physiology through other mechanisms. Native ellagitannin forms of PGPE are hydrolysed by intestinal microflora into punicalagins and ellagic acid, which act as prebiotics. These molecules might contribute to the positive effect on bone [75,76]. They are known to enhance the growth of bifidobacteria and lactobacilli in the intestine, leading to increased production of short chain fatty acids (SCFAs) [77]. SCFAs contribute directly to the enhancement of Ca absorption via a cation exchange mechanism (increased exchange of cellular H^+ for luminal Ca^{2+}) [78,79]. The extract of pomegranate pericarp could act as selective oestrogen receptor modulators (SERMs) on bone as well, as it has an antioestrogenic effect in the mammary gland, without compromising the beneficial effects of oestrogen in the cardiovascular and skeletal system [80]. This SERM-like effect could be attributed to ellagic acid. *In vivo*, this compound prevents bone loss after tooth extraction in normal [81] or diabetic rats [82] with an increase of ALP expression. In ovariectomized (OVX) rats, an extract of Punica granatum peel has been shown to induce a simultaneous, dose-dependent increase in femoral BMD and uterine weight, suggesting a potent oestrogenic activity of the ellagic acid present in the extract [83]. These effects were equivalent to the effects of Tamoxifen in increasing femoral mineral content and osteoblast number [84]. Moreover, OVX rats fed with dried pomegranate concentrate powder (PCP) extracts containing 0.90 mg/g of ellagic acid show increased serum estradiol and bALP levels with decreased serum osteocalcin compared with OVX control rats [85]. The failure load of the femur was also significantly increased, suggesting that the PCP treatment activates osteoblast differentiation and inhibits bone mineralization and turnover. *In vitro*, ellagic acid induces nodule mineralization in an osteoblastic cell line (KS483) through a pathway involving the oestrogen receptor [86,87]. However in our *in vivo* experiment, we did not observe any difference in uterine weight between the OVX control and the OVX PGPE groups. This result could be explained by the lower dose [83].

In addition, looking beyond the traditional bone perspective, because inflammation and oxidative stress play a major role in the modulation of bone remodelling, [47,88], we investigated the expression of major markers of such processes using Taqman Low density Arrays. We showed that PGPE consumption lowers pro-inflammatory markers and enzymes implicated in reactive oxygen species, ROS synthesis expression (Chemokine ligand 2, CCL2; Interleukin 1 receptor 1, IL1-R1; nitric oxide synthase 2, NOS2 and NADPH oxidase 4, NOX4) and enhances the expression of anti-inflammatory markers and the enzymes implicated in the anti-oxidant process (Interleukin 1 receptor antagonist, IL1-Rn; glutathione peroxidase, GPx; glutathione reductase, GSR). The flavonoid content and their derivatives (rutin, gallic acid, ellagic acid and punicalagin) of the peel fraction of pomegranate have been frequently utilized as antioxidants in various dietary supplements [76,89]. *In vivo*, a pomegranate extract diet rich in ellagitannins has been acknowledged to suppress inflammation by decreasing the IL-6 level in the joints of collagen-induced arthritis in mice [90], while the consumption of pomegranate decreases oxidative status in patients with rheumatoid arthritis [91]. *In vitro*, it has been shown that the hydrolysable tannin punicalagin from dried pomegranate peels may have an anti-inflammatory effect on bone metabolism by inhibiting PGE2 production and cyclooxygenase 2, COX-2 expression in lipopolysaccharide, LPS-induced RAW 264.7 cells [92] and/or repressing osteoclast differentiation through the downregulation of Nuclear Factor Of Activated T-Cells, NFATc1 expression and decreased phosphorylation of the Akt and JNK pathways [93]. Likewise, the anti-inflammatory and antioxidant effects of urolithins have been extensively demonstrated *in vitro* [94,95].

In summary, our study demonstrates that PGPE metabolites can directly modulate bone cell differentiation, leading to an improved resorption/formation ratio together with anti-inflammatory and anti-oxidative effects in the bone microenvironment. This ability can explain the improvement in bone mass and microarchitecture in our model of osteoporosis. These encouraging data suggest that pomegranate consumption could be a promising alternative and complementary therapeutic agent for the prevention of osteoporosis. Nevertheless, more studies are still warranted to further determine the molecular mechanisms involved and to investigate whether these *ex vivo* and preclinical data can be extrapolated to the human situation.

ACKNOWLEDGMENTS

Financial support was provided by Greentech and INRA, UMR 1019, UNH, Clermont-Ferrand. The authors are also grateful to the staff from the Animal Facilities (Installation Experimentale de Nutrition, Saint Genès Champanelle, France [56]) who provided everyday care to the mice.

AUTHOR CONTRIBUTIONS

M.S., L.R., Y.W. and V.C. contributed to the study conception and design. M.S., L.L., M.J.D. and P.L. collected data and performed analysis. P.P. and E.M.N. performed the µCt analysis. M.S., Y.W. and V.C. interpreted the data. M.S., L.L., M.J.D., Y.W. and V.C. drafted the manuscript. All the authors critically revised and approved its final version.

CONFLICT OF INTEREST

Mélanie Spilmont was a PhD student funded by Greentech. Laurent Rios worked for Greentech during this project and is now a university professor. Laurent Léotoing, Marie-Jeanne Davicco, Patrice Lebecque, Elisabeth Miot-Noirault, Paul Pilet, Yohann Wittrant, and Véronique Coxam declare that they have no competing interests.

REFERENCES

1. Rachner, T.D.; Khosla, S.; Hofbauer, L.C. Osteoporosis: Now and the future. *Lancet* **2011**, *377*, 1276–1287.

2. Sandhu, S.K.; Hampson, G. The pathogenesis, diagnosis, investigation and management of osteoporosis. *J. Clin. Pathol.* **2011**, *64*, 1042–1050.

3. Schulman, R.C.; Weiss, A.J.; Mechanick, J.I. Nutrition, bone, and aging: An integrative physiology approach. *Curr. Osteoporos. Rep.* **2011**, *9*, 184–195.

4. Scalbert, A.; Manach, C.; Morand, C.; Remesy, C.; Jimenez, L. Dietary polyphenols and the prevention of diseases. *Crit. Rev. Food Sci. Nutr.* **2005**, *45*, 287–306.

5. Fonseca, D.; Ward, W.E. Detection of isoflavones in mouse tibia after feeding daidzein. *J. Med. Food* **2006**, *9*, 436–439.

6. Hagiwara, K.; Goto, T.; Araki, M.; Miyazaki, H.; Hagiwara, H. Olive polyphenol hydroxytyrosol prevents bone loss. *Eur. J. Pharmacol.* **2011**, *662*, 78–84.

7. Ishimi, Y.; Yoshida, M.; Wakimoto, S.; Wu, J.; Chiba, H.; Wang, X.; Takeda, K.; Miyaura, C. Genistein, a soybean isoflavone, affects bone marrow lymphopoiesis and prevents bone loss in castrated male mice. *Bone* **2002**, *31*, 180–185

8. Puel, C.; Mathey, J.; Agalias, A.; Kati-Coulibaly, S.; Mardon, J.; Obled, C.; Davicco, M.J.; Lebecque, P.; Horcajada, M.N.; Skaltsounis, A.L.; *et al.* Dose-response study of effect of oleuropein, an olive oil polyphenol, in an ovariectomy/inflammation experimental model of bone loss in the rat. *Clin. Nutr.* **2006**, *25*, 859–868.

9. Wu, J.; Wang, X.X.; Chiba, H.; Higuchi, M.; Takasaki, M.; Ohta, A.; Ishimi, Y. Combined intervention of exercise and genistein prevented androgen deficiency-induced bone loss in mice.*J. Appl. Physiol.* **2003**, *94*, 335–342.

10. Horcajada, M.N.; Offord, E. Naturally plant-derived compounds: Role in bone anabolism. *Curr. Mol. Pharmacol.* **2012**, *5*, 205–218.

11. Li, B.; Yu, S. Genistein prevents bone resorption diseases by inhibiting bone resorption and stimulating bone formation. *Biol. Pharm. Bull.* **2003**, *26*, 780–786.

12. Rassi, C.M.; Lieberherr, M.; Chaumaz, G.; Pointillart, A.; Cournot, G. Down-regulation of osteoclast differentiation by daidzein via caspase 3. *J. Bone Miner. Res.* **2002**, *17*, 630–638.

13. Trzeciakiewicz, A.; Habauzit, V.; Horcajada, M.N. When nutrition interacts with osteoblast function: Molecular mechanisms of polyphenols. *Nutr. Res. Rev.* **2009**, *22*, 68–81.

14. De Wilde, A.; Lieberherr, M.; Colin, C.; Pointillart, A. A low dose of daidzein acts as an erbeta-selective agonist in trabecular osteoblasts of young female piglets. *J. Cell Physiol.* **2004**, *200*, 253–262.

15. Santiago-Mora, R.; Casado-Diaz, A.; de Castro, M.D.; Quesada-Gomez, J.M. Oleuropein enhances osteoblastogenesis and inhibits adipogenesis: The effect on differentiation in stem cells derived from bone marrow. *Osteoporos. Int.* **2011**, *22*, 675–684.

16. Sugimoto, E.; Yamaguchi, M. Anabolic effect of genistein in osteoblastic MC3T3-E1 cells. *Int. J. Mol. Med.* **2000**, *5*, 515–520.

17. Lansky, E.P.; Newman, R.A. *Punica granatum* (pomegranate) and its potential for prevention and treatment of inflammation and cancer. *J. Ethnopharmacol.* **2007**, *109*, 177–206.

18. Viuda-Martos, M.; Fernández-López, J.; Pérez-Álvarez, J.A. Pomegranate and its many functional components as related to human health: A review. *Compr. Rev. Food Sci. Food Saf.* **2010**, *9*, 635–654.

19. Rahimi, H.R.; Arastoo, M.; Ostad, S.N. A comprehensive review of *punica granatum* (pomegranate) properties in toxicological, pharmacological, cellular and molecular biology researches. *Iran J. Pharm. Res.* **2012**, *11*, 385–400.

20. Kim, Y.H.; Choi, E.M. Stimulation of osteoblastic differentiation and inhibition of interleukin-6 and nitric oxide in Mc3t3-E1 cells by pomegranate ethanol extract. *Phytother. Res.* **2009**, *23*, 737–739.

21. Rosenblat, M.; Volkova, N.; Aviram, M. Pomegranate juice (PJ) consumption antioxidative properties on mouse macrophages, but not pj beneficial effects on macrophage cholesterol and triglyceride metabolism, are mediated via pj-induced stimulation of macrophage pon2.*Atherosclerosis* **2010**, *212*, 86–92.

22. Hollebeeck, S.; Winand, J.; Herent, M.F.; During, A.; Leclercq, J.; Larondelle, Y.; Schneider, Y.J. Anti-inflammatory effects of pomegranate (*Punica granatum* L.) husk ellagitannins in Caco-2 cells, an *in vitro* model of human intestine. *Food Funct.* **2012**, *3*, 875–885.

23. Larrosa, M.; Gonzalez-Sarrias, A.; Yanez-Gascon, M.J.; Selma, M.V.; Azorin-Ortuno, M.; Toti, S.; Tomas-Barberan, F.; Dolara, P.; Espin, J.C. Anti-inflammatory properties of a pomegranate extract and its metabolite urolithin-A in a

colitis rat model and the effect of colon inflammation on phenolic metabolism. *J. Nutr. Biochem.* **2010**, *21*, 717–725.

24. Kasimsetty, S.G.; Bialonska, D.; Reddy, M.K.; Ma, G.; Khan, S.I.; Ferreira, D. Colon cancer chemopreventive activities of pomegranate ellagitannins and urolithins. *J. Agric. Food Chem.* **2010**,*58*, 2180–2187.

25. Larrosa, M.; Tomas-Barberan, F.A.; Espin, J.C. The dietary hydrolysable tannin punicalagin releases ellagic acid that induces apoptosis in human colon adenocarcinoma Caco-2 cells by using the mitochondrial pathway. *J. Nutr. Biochem.* **2006**, *17*, 611–625.

26. Syed, D.N.; Chamcheu, J.C.; Mukhtar, V.M. Pomegranate extracts and cancer prevention: Molecular and cellular activities. *Anticancer Agents Med. Chem.* **2012**, *13*, 1149–1161.

27. Aviram, M.; Volkova, N.; Coleman, R.; Dreher, M.; Reddy, M.K.; Ferreira, D.; Rosenblat, M. Pomegranate phenolics from the peels, arils, and flowers are antiatherogenic: Studies *in vivo* in atherosclerotic apolipoprotein e-deficient (e 0) mice and *in vitro* in cultured macrophages and lipoproteins. *J. Agric. Food Chem.* **2008**, *56*, 1148–1157.

28. Rosenblat, M.; Volkova, N.; Coleman, R.; Aviram, M. Pomegranate byproduct administration to apolipoprotein e-deficient mice attenuates atherosclerosis development as a result of decreased macrophage oxidative stress and reduced cellular uptake of oxidized low-density lipoprotein. *J. Agric. Food Chem.* **2006**, *54*, 1928–1935.

29. Fuhrman, B.; Volkova, N.; Aviram, M. Pomegranate juice inhibits oxidized LDL uptake and cholesterol biosynthesis in macrophages. *J. Nutr. Biochem.* **2005**, *16*, 570–576.

30. Esmaillzadeh, A.; Tahbaz, F.; Gaieni, I.; Alavi-Majd, H.; Azadbakht, L. Concentrated pomegranate juice improves lipid profiles in diabetic patients with hyperlipidemia. *J. Med. Food* **2004**, *7*, 305–308.

31. Gonzalez-Ortiz, M.; Martinez-Abundis, E.; Espinel-Bermudez, M.C.; Perez-Rubio, K.G. Effect of pomegranate juice on insulin secretion and sensitivity in patients with obesity. *Ann. Nutr. Metab.***2011**, *58*, 220–223.

32. Bagri, P.; Ali, M.; Aeri, V.; Bhowmik, M.; Sultana, S. Antidiabetic effect of punica granatum flowers: Effect on hyperlipidemia, pancreatic cells lipid peroxidation and antioxidant enzymes in experimental diabetes. *Food Chem. Toxicol.* **2009**, *47*, 50–54.

33. Fawole, O.A.; Makunga, N.P.; Opara, U.L. Antibacterial, antioxidant and tyrosinase-inhibition activities of pomegranate fruit peel methanolic extract. *BMC Complement Altern. Med.* **2012**, *12*, 200.

34. Abdollahzadeh, S.; Mashouf, R.; Mortazavi, H.; Moghaddam, M.; Roozbahani, N.; Vahedi, M. Antibacterial and antifungal activities of *Punica granatum* peel extracts against oral pathogens. *J. Dent.* **2011**, *8*, 1–6.

35. Haidari, M.; Ali, M.; Ward Casscells, S., 3rd; Madjid, M. Pomegranate (*punica granatum*) purified polyphenol extract inhibits influenza virus and has a synergistic effect with oseltamivir.*Phytomedicine* **2009**, *16*, 1127–1136.

36. Johanningsmeier, S.D.; Harris, G.K. Pomegranate as a functional food and nutraceutical source.*Annu. Rev. Food Sci. Technol.* **2011**, *2*, 181–201.

37. Haber, S.L.; Joy, J.K.; Largent, R. Antioxidant and antiatherogenic effects of pomegranate. *Am. J. Health Syst. Pharm.* **2011**, *68*, 1302–1305.

38. Guo, C.; Yang, J.; Wei, J.; Li, Y.; Xu, J.; Jiang, Y. Antioxidant activities of peel, pulp and seed fractions of common fruits as determined by frap assay. *Nutr. Res.* **2003**, *23*, 1719–1726.

39. Noda, Y.; Kaneyuki, T.; Mori, A.; Packer, L. Antioxidant activities of pomegranate fruit extract and its anthocyanidins: Delphinidin, cyanidin, and pelargonidin. *J. Agric. Food Chem.* **2002**, *50*, 166–171.

40. Krause, K.H. Aging: A revisited theory based on free radicals generated by NOX family NADPH oxidases. *Exp. Gerontol.* **2007**, *42*, 256–262.

41. Probst-Hensch, N.M. Chronic age-related diseases share risk factors: Do they share pathophysiological mechanisms and why does that matter? *Swiss Med. Wkly.* **2010**.

42. Syslova, K.; Bohmova, A.; Mikoska, M.; Kuzma, M.; Pelclova, D.; Kacer, P. Multimarker screening of oxidative stress in aging. *Oxid. Med Cell. Longev.* **2014**, *2014*, 562–860.

43. Balcerczyk, A.; Gajewska, A.; Macierzynska-Piotrowska, E.; Pawelczyk, T.; Bartosz, G.; Szemraj, J. Enhanced antioxidant capacity and anti-ageing biomarkers after diet micronutrient supplementation. *Molecules* **2014**, *19*, 14794–14808.

44. Howcroft, T.K.; Campisi, J.; Louis, G.B.; Smith, M.T.; Wise, B.; Wyss-Coray, T.; Augustine, A.D.; McElhaney, J.E.; Kohanski, R.; Sierra, F. The role of inflammation in age-related disease. *Aging (Albany NY)* **2013**, *5*, 84–93.

45. Wauquier, F.; Barquissau, V.; Leotoing, L.; Davicco, M.J.; Lebecque, P.; Mercier, S.; Philippe, C.; Miot-Noirault, E.; Chardigny, J.M.; Morio, B.; *et al.* Borage and fish oils lifelong supplementation decreases inflammation and improves

bone health in a murine model of senile osteoporosis. *Bone***2012**, *50*, 553–561.

46. Lacativa, P.G.; Farias, M.L. Osteoporosis and inflammation. *Arq Bras Endocrinol. Metabol.* **2010**, *54*, 123–132.

47. Wauquier, F.; Leotoing, L.; Coxam, V.; Guicheux, J.; Wittrant, Y. Oxidative stress in bone remodelling and disease. *Trends Mol. Med.* **2009**, *15*, 468–477.

48. Manolagas, S.C.; Parfitt, A.M. What old means to bone. *Trends Endocrinol. Metab.* **2010**, *21*, 369–374.

49. Koh, J.M.; Lee, Y.S.; Kim, Y.S.; Kim, D.J.; Kim, H.H.; Park, J.Y.; Lee, K.U.; Kim, G.S. Homocysteine enhances bone resorption by stimulation of osteoclast formation and activity through increased intracellular ros generation. *J. Bone Miner. Res.* **2006**, *21*, 1003–1011.

50. Ozgocmen, S.; Kaya, H.; Fadillioglu, E.; Aydogan, R.; Yilmaz, Z. Role of antioxidant systems, lipid peroxidation, and nitric oxide in postmenopausal osteoporosis. *Mol. Cell. Biochem.* **2007**, *295*, 45–52.

51. Zahin, M.; Aqil, F.; Ahmad, I. Broad spectrum antimutagenic activity of antioxidant active fraction of *Punica granatum* L. Peel extracts. *Mutat. Res.* **2010**, *703*, 99–107.

52. Tzulker, R.; Glazer, I.; Bar-Ilan, I.; Holland, D.; Aviram, M.; Amir, R. Antioxidant activity, polyphenol content, and related compounds in different fruit juices and homogenates prepared from 29 different pomegranate accessions. *J. Agric. Food Chem.* **2007**, *55*, 9559–9570.

53. Ismail, T.; Sestili, P.; Akhtar, S. Pomegranate peel and fruit extracts: A review of potential anti-inflammatory and anti-infective effects. *J. Ethnopharmacol.* **2012**, *143*, 397–405.

54. Mousavinejad, G.; Emam-Djomeh, Z.; Rezaei, K.; Khodaparast, M.H.H. Identification and quantification of phenolic compounds and their effects on antioxidant activity in pomegranate juices of eight iranian cultivars. *Food Chem.* **2009**, *115*, 1274–1278.

55. Manach, C.; Scalbert, A.; Morand, C.; Remesy, C.; Jimenez, L. Polyphenols: Food sources and bioavailability. *Am. J. Clin. Nutr.* **2004**, *79*, 727–747.

56. Installation expérimentale de nutrition. Available online: https://www6.clermont.inra.fr/unh/PlateauTechniques/Installation-Experimentale-de-Nutrition (accessed on 26 October 2015).

57. Sabokbar, A.; Millett, P.J.; Myer, B.; Rushton, N. A rapid, quantitative assay for measuring alkaline phosphatase activity in osteoblastic cells *in vitro*. *Bone Miner.* **1994**, *27*, 57–67.

58. Trzeciakiewicz, A.; Habauzit, V.; Mercier, S.; Barron, D.; Urpi-Sarda, M.; Manach, C.; Offord, E.; Horcajada, M.N. Molecular mechanism of hesperetin-*7-O*-glucuronide, the main circulating metabolite of hesperidin, involved in osteoblast differentiation. *J. Agric. Food Chem.* **2010**, *58*, 668–675.

59. Wittrant, Y.; Gorin, Y.; Woodruff, K.; Horn, D.; Abboud, H.E.; Mohan, S.; Abboud-Werner, S.L. High D(+)glucose concentration inhibits rankl-induced osteoclastogenesis. *Bone* **2008**, *42*, 1122–1130.

60. Kalu, D.N. The ovariectomized rat model of postmenopausal bone loss. *Bone Miner.* **1991**, *15*, 175–191.

61. Lei, Z.; Xiaoying, Z.; Xingguo, L. Ovariectomy-associated changes in bone mineral density and bone marrow

haematopoiesis in rats. *Int. J. Exp. Pathol.* **2009**, *90*, 512–519.

62. Bruker. Available online: http://bruker-microct.com/products/downloads.htm (accessed on 26 October 2015).

63. XLSTAT. Available online: https://www.xlstat.com/fr/ (accessed on 26 October 2015).

64. Espin, J.C.; Gonzalez-Barrio, R.; Cerda, B.; Lopez-Bote, C.; Rey, A.I.; Tomas-Barberan, F.A. Iberian pig as a model to clarify obscure points in the bioavailability and metabolism of ellagitannins in humans. *J. Agric. Food Chem.* **2007**, *55*, 10476–10485.

65. Gonzalez-Sarrias, A.; Gimenez-Bastida, J.A.; Garcia-Conesa, M.T.; Gomez-Sanchez, M.B.; Garcia-Talavera, N.V.; Gil-Izquierdo, A.; Sanchez-Alvarez, C.; Fontana-Compiano, L.O.; Morga-Egea, J.P.; Pastor-Quirante, F.A.; *et al.* Occurrence of urolithins, gut microbiota ellagic acid metabolites and proliferation markers expression response in the human prostate gland upon consumption of walnuts and pomegranate juice. *Mol. Nutr. Food Res.* **2010**, *54*, 311–322.

66. Patel, C.; Dadhaniya, P.; Hingorani, L.; Soni, M.G. Safety assessment of pomegranate fruit extract: Acute and subchronic toxicity studies. *Food Chem. Toxicol.* **2008**, *46*, 2728–2735.

67. Zhang, C. Molecular mechanisms of osteoblast-specific transcription factor osterix effect on bone formation. *Beijing Da Xue Xue Bao* **2012**, *44*, 659–665.

68. Zhang, Y.; Su, J.; Yu, J.; Bu, X.; Ren, T.; Liu, X.; Yao, L. An essential role of discoidin domain receptor 2 (DDR2) in osteoblast differentiation and chondrocyte maturation via

modulation of Runx2 activation. *J. Bone Miner. Res.* **2011**, *26*, 604–617.

69. Sreekumar, S.; Sithul, H.; Muraleedharan, P.; Azeez, J.M.; Sreeharshan, S. Pomegranate fruit as a rich source of biologically active compounds. *BioMed. Res. Int.* **2014**, *2014*, 686921.

70. Boyle, W.J.; Simonet, W.S.; Lacey, D.L. Osteoclast differentiation and activation. *Nature* **2003**, *423*, 337–342.

71. Vaira, S.; Johnson, T.; Hirbe, A.C.; Alhawagri, M.; Anwisye, I.; Sammut, B.; O'Neal, J.; Zou, W.; Weilbaecher, K.N.; Faccio, R.; *et al.* RelB is the NF-kB subunit downstream of nik responsible for osteoclast differentiation. *Proc. Natl. Acad. Sci. USA* **2008**, *105*, 3897–3902.

72. Coelho, L.F.; Magno de Freitas Almeida, G.; Mennechet, F.J.; Blangy, A.; Uze, G. Interferon-alpha and -beta differentially regulate osteoclastogenesis: Role of differential induction of chemokine cxcl11 expression. *Proc. Natl. Acad. Sci. USA* **2005**, *102*, 11917–11922.

73. Lin, Y.; Murray, M.A.; Garrett, I.R.; Gutierrez, G.E.; Nyman, J.S.; Mundy, G.; Fast, D.; Gellenbeck, K.W.; Chandra, A.; Ramakrishnan, S. A targeted approach for evaluating preclinical activity of botanical extracts for support of bone health. *J. Nutr. Sci.* **2014**, *3*, e13.

74. Park, E.K.; Kim, M.S.; Lee, S.H.; Kim, K.H.; Park, J.Y.; Kim, T.H.; Lee, I.S.; Woo, J.T.; Jung, J.C.; Shin, H.I.; *et al.* Furosin, an ellagitannin, suppresses rankl-induced osteoclast differentiation and function through inhibition of map kinase activation and actin ring formation. *Biochem. Biophys. Res. Commun.* **2004**, *325*, 1472–1480.

75. Li, Z.; Summanen, P.H.; Komoriya, T.; Henning, S.M.; Lee, R.P.; Carlson, E.; Heber, D.; Finegold, S.M. Pomegranate

ellagitannins stimulate growth of gut bacteria *in vitro*: Implications for prebiotic and metabolic effects. *Anaerobe* **2015**, *34*, 164–168.

76. Akhtar, S.; Ismail, T.; Fraternale, D.; Sestili, P. Pomegranate peel and peel extracts: Chemistry and food features. *Food Chem.* **2015**, *174*, 417–425.

77. Bialonska, D.; Ramnani, P.; Kasimsetty, S.G.; Muntha, K.R.; Gibson, G.R.; Ferreira, D. The influence of pomegranate by-product and punicalagins on selected groups of human intestinal microbiota. *Int. J. Food Microbiol.* **2010**, *140*, 175–182.

78. Lutz, T.; Scharrer, E. Effect of short-chain fatty acids on calcium absorption by the rat colon. *Exp. Physiol.* **1991**, *76*, 615–618.

79. Roberfroid, M.; Gibson, G.R.; Hoyles, L.; McCartney, A.L.; Rastall, R.; Rowland, I.; Wolvers, D.; Watzl, B.; Szajewska, H.; Stahl, B.; *et al.* Prebiotic effects: Metabolic and health benefits. *Br. J. Nutr.***2010**, *104*, S1–S63.

80. Sreeja, S.; Santhosh Kumar, T.R.; Lakshmi, B.S. Pomegranate extract demonstrate a selective estrogen receptor modulator profile in human tumor cell lines and *in vivo* models of estrogen deprivation. *J. Nutr. Biochem.* **2012**, *23*, 725–732.

81. Al-Obaidi, M.M.; Al-Bayaty, F.H.; Al Batran, R.; Hassandarvish, P.; Rouhollahi, E. Protective effect of ellagic acid on healing alveolar bone after tooth extraction in rat—A histological and immunohistochemical study. *Arch. Oral Biol.* **2014**, *59*, 987–999.

82. Al-Obaidi, M.M.; Al-Bayaty, F.H.; Al Batran, R.; Hussaini, J.; Khor, G.H. Impact of ellagic acid in bone formation after

tooth extraction: An experimental study on diabetic rats. *Sci. World J.* **2014**,*2014*, 908098.

83. Satpathy, S.; Patra, A.; Purohit, A.P. Estrogenic activity of *Punica granatum* L. Pell estract. *Asian Pac. J. Reprod.* **2013**, *2*, 19–24.

84. Bahtiar, A.; Arifin, S.; Razalifha, A.; Qomariah, N.; Wuyung, P.; Arsianti, A. Polar fraction of*Punica granatum* L. Peel extract increased osteoblast number on ovariectomized rat bone. *Int. J. Herbal Med.* **2014**, *2*, 65–70.

85. Kang, S.J.; Choi, B.R.; Kim, S.H.; Yi, H.Y.; Park, H.R.; Kim, D.C.; Choi, S.H.; Han, C.H.; Park, S.J.; Song, C.H.; *et al.* Dried pomegranate potentiates anti-osteoporotic and anti-obesity activities of red clover dry extracts in ovariectomized rats. *Nutrients* **2015**, *7*, 2622–2647.

86. Papoutsi, Z.; Kassi, E.; Chinou, I.; Halabalaki, M.; Skaltsounis, L.A.; Moutsatsou, P. Walnut extract (juglans regia l.) and its component ellagic acid exhibit anti-inflammatory activity in human aorta endothelial cells and osteoblastic activity in the cell line ks483. *Br. J. Nutr.* **2008**, *99*, 715–722.

87. Papoutsi, Z.; Kassi, E.; Tsiapara, A.; Fokialakis, N.; Chrousos, G.P.; Moutsatsou, P. Evaluation of estrogenic/antiestrogenic activity of ellagic acid via the estrogen receptor subtypes eralpha and erbeta. *J. Agric. Food Chem.* **2005**, *53*, 7715–7720.

88. Mundy, G.R. Osteoporosis and inflammation. *Nutr. Rev.* **2007**, *65*, S147–S151.

89. Middha, S.K.; Usha, T.; Pande, V. HPLC evaluation of phenolic profile, nutritive content, and antioxidant capacity of extracts obtained from punica granatum fruit peel. *Adv. Pharmacol. Sci.***2013**, *2013*, 296236.

90. Shukla, M.; Gupta, K.; Rasheed, Z.; Khan, K.A.; Haqqi, T.M. Consumption of hydrolyzable tannins-rich pomegranate extract suppresses inflammation and joint damage in rheumatoid arthritis. *Nutrition* **2008**, *24*, 733–743.

91. Balbir-Gurman, A.; Fuhrman, B.; Braun-Moscovici, Y.; Markovits, D.; Aviram, M. Consumption of pomegranate decreases serum oxidative stress and reduces disease activity in patients with active rheumatoid arthritis: A pilot study. *Isr. Med. Assoc. J.* **2011**, *13*, 474–479.

92. Lee, C.J.; Chen, L.G.; Liang, W.L.; Wanga, C. Anti-inflammatory effects of punica granatum linne*in vitro* and *in vivo*. *Food Chem.* **2010**, *118*, 315–322.

93. Iwatake, M.; Okamoto, K.; Tanaka, T.; Tsukuba, T. Punicalagin attenuates osteoclast differentiation by impairing nfatc1 expression and blocking AKT- and JNK-dependent pathways. *Mol. Cell. Biochem.* **2015**, *407*, 161–172.

94. Bialonska, D.; Kasimsetty, S.G.; Khan, S.I.; Ferreira, D. Urolithins, intestinal microbial metabolites of pomegranate ellagitannins, exhibit potent antioxidant activity in a cell-based assay. *J. Agric. Food Chem.* **2009**, *57*, 10181–10186.

95. Gonzalez-Sarrias, A.; Larrosa, M.; Tomas-Barberan, F.A.; Dolara, P.; Espin, J.C. NF-κB-dependent anti-inflammatory activity of urolithins, gut microbiota ellagic acid-derived metabolites, in human colonic fibroblasts. *Br. J. Nutr.* **2010**, *104*, 503–512.

10

CHAPTER

Effects of Albedo Addition on Pomegranate Juice Physicochemical, Volatile and Chemical Markers

Laura Vázquez-Araújo [1,*], Edgar Chambers IV [1] and Ángel A. Carbonell-Barrachina [2]

[1] *The Sensory Analysis Center, Kansas State University, 1310 Research Park Drive, Manhattan, KS 66502, USA*

[2] *Food Quality and Safety Group, Agro-Food Technology Department, Universidad Miguel Hernández de Elche, Escuela Politécnica Superior de Orihuela, Carretera de Beniel, km 3.2, 03312-Orihuela, Alicante, Spain*

ABSTRACT

Five commercial juices, representing the five clusters of this juice, were characterized before and after maceration with 10% pomegranate albedo (control- and albedo treated (AT)-juices, respectively). Commercial juices were macerated with albedo homogenate for 24 h, and then the albedo was removed. Total soluble solids, titratable acidity, maturity index (MI), total phenolic content (TPC), volatile composition, and flavor profile were evaluate in control- and AT-juices. From all physico-chemical characteristics, only the TPC was significantly affected by the treatment and ranged from 846 to 3784 mg gallic acid L^{-1} and from 2163 to 5072 mg gallic acid L^{-1} in control- and AT-juices, respectively; the increment in TPC was more than 1.3-fold in all

AT-juices. No clear pattern was found when studying the volatile composition; only significant increases were observed in the contents of hexanal, 2-hexenal, and 3-hexenal in all AT-samples. The flavor profile study indicated that three of the five samples increased their bitterness and/or astringency. In addition, new attributes, which were not present in the control juices, appeared after maceration with albedo in some samples: green-bean, brown-sweet, and green-viney. This information will be useful in developing and promoting new "healthy" products based on pomegranate.

Keywords: *Punica granatum* L.; flavor; phenolic content; SPME-GCMS; volatile compounds

1. INTRODUCTION

Pomegranate and its derivatives, such as pomegranate juice, have become very popular in recent years. Pomegranate has been reported as being capable of addressing different health diseases, or at least having a significant effect over them. This fruit, and especially its juice, seems to have, among others, antiatherogenic, antioxidant, and antihypertensive effects [1,2,3]. As reported by Johanningsmeier and Harris [4], the sales of pomegranate juice increased from over $84,500 in 2001 to $66 million in 2005 in the USA, probably due to the wide promotion of its healthy effects.

The antioxidant activity, AA, of pomegranate juice is positively correlated with its total phenolic content, TPC [5,6,7]. The phenolic compounds are more abundant in the non-edible portions of pomegranate, especially the rind and carpelar membranes [8,9]. Consequently, it is reasonable to assume that depending on the extraction method used to obtain the juice, the final product will have different antioxidant activity. In this way, processing steps extending the maceration of the juice with rind will lead to juices with high values of TPC. The TPC has been studied previously by

Gil *et al.* [8] and Tezcan*et al.* [9] in commercial juices, and large differences were found. Recently, other authors [10,11] have determined that the cultivar used to elaborate the juice has also a significant effect on TPC and the associated AA.

Ibrahim [12] reported that pomegranate rind extract had strong antimicrobial effect, high AA, and also enhanced liver and kidney functions in animal models. Therefore, this type of rind extract can be used as food preservative or even as a nutraceutical ingredient for new enriched foods [13]. However, there are no scientific studies describing the effects of the addition of pomegranate extracts on the chemical and/or sensory quality of pomegranate juices.

Koppel and Chambers [14] studied 33 commercial pomegranate juices and developed a sensory lexicon to describe the main sensory attributes of these products. As a result of their study, pomegranate juices were grouped into five different clusters characterized by the following sensory attributes: *cluster 1* berry, dark-fruity flavors, and toothetch; *cluster 2* grape, cranberry, and wine-like flavor; *cluster 3*fermented flavor and toothetch; *cluster 4* brown color, and a characteristic musty/earthy flavor; and*cluster 5* candy-like and sweet overall flavors.

The aim of this study was to determine the influence of macerating pomegranate albedo with pomegranate juice on the main chemical and sensory characteristics of the juice. To achieve this goal, the main physico-chemical (total soluble solid content, titratable acidity, maturity index, and total phenolic content), aromatic (volatile composition), and flavor (sensory profile) characteristics of five commercial juices, which represented each one of the aforementioned clusters, were studied before and after macerating them with a 10 % of pomegranate albedo.

2. MATERIALS AND METHODS

2.1. Samples

Three different pomegranate juices were purchased from different parts of the US and shipped to the Sensory Analysis Center (Kansas State University), in Manhattan, KS, USA. The fourth and fifth juices were purchased in Estonia and Spain, respectively, and shipped in the same way to Kansas. These five samples had been previously studied by Koppel and Chambers [14] and were chosen in the present study for representing the five different aforementioned clusters: 618 (cluster 1), 324 (cluster 2), 707 (cluster 3), 612 (cluster 4), and 981 (cluster 5).

At the same time, 15 pomegranates, cultivar *Wonderful*, were purchased from a local grocery store in Manhattan, KS. After discarding damaged fruits, the arils from all the fruits were manually removed, and the albedo and carpelar membranes of each pomegranate rind were grated, blended in a food processor, and frozen (−20 °C) until preparation of albedo-treated (AT-juice) samples.

Initially, the physico-chemical parameters and the volatile composition were analyzed in control juices. Later, 10% of homogenized pomegranate albedo was added to the control juices to prepare AT-juices. Samples were left macerating at 4 °C during 24 h; then, the juices were filtered using a strainer (mesh size < 1 mm) to remove all solid particles. Three batches of AT-juices were prepared, and all samples were analyzed in triplicate.

2.1. Physico-Chemical Analysis

2.1.1. Total Soluble Solids, Titratable Acidity, and pH

Total soluble solids (TSS) were measured with a digital refractometer (Model PR-101a; Atago, Bellevue, WA, USA) at ~20 °C, with values being expressed as °Brix. Titratable acidity (TA) was determined by titrating 1 mL of each sample (diluted to 20 mL final volume with deionized water) with 0.1 mol L^{-1} NaOH. Results were expressed as g citric acid 100 mL^{-1}. pH was measured with a pH-meter (Accumet Basic AB15, Thermo Fisher Scientific, Waltham, MA, USA). All analyses were run in triplicate, with each replication corresponding to a different bottle of juice. Finally, the maturity index (MI: ratio of TSS to TA) was calculated for each sample.

2.1.2. Total Phenolic Content

Total phenolic content (TPC) was measured as indicator of the antioxidant activity in the juices. TPC was determined by using the Folin-Ciocalteau method with some modifications [15]. Results were expressed as mg gallic acid equivalent L^{-1}. Analyses were run in triplicate.

2.2. Analysis of Volatile Composition

2.2.1. Extraction Procedure

Two mL of each sample were placed in a 10 mL vial with a polypropylene hole cap PTFE/silicone septa. The compound 1,2-dimethoxybenzene was used as internal standard to semi-quantify the volatile compounds. The vials were equilibrated during 10 min at 60 °C in the autosampler (Pal system, model CombiPal, CTC Analytics AG, Zwingen, Switzerland). After this equilibration time, a 50/30 μm DVB/CAR/PDMS fiber was exposed to the sample headspace for 30 min at 60 °C. The desorption of the volatile compounds from the fiber coating was made in the injection port of GC at 250 °C during 5 min in splitless mode. Experiments were run in triplicate.

2.2.2. Chromatographic Analyses

The isolation, identification, and quantification of the volatile compounds were performed on a gas chromatograph (Varian GC CP3800; Varian, Inc., Walnut Creek, CA, USA), coupled with a Varian mass spectrometer detector (Saturn 2200), and operated with the MS Workstation software. The GC-MS system was equipped with a VF-5MS column (Varian, Inc., Walnut Creek, CA; 30 m × 0.25 mm × 1.0 μm film thickness). The temperature of the column began at 40 °C, was held for 10 min, increased 8 °C min^{-1} to 180 °C, and, finally increased 10 °C min^{-1} to 280 °C, and held for 10 min. The column flow was 1 mL min^{-1}, using helium as the carrier gas.

Most compounds were identified using two different analytical methods: (1) retention indexes, and (2) mass spectra (authentic chemicals and Wiley spectral library collection).

2.3. Sensory Evaluation with Trained Panel

Six highly trained panelists from the Sensory Analysis Center participated in this study. Each panelist had more than 120 h of training in sensory testing and more than 1000 h of testing experience with a variety of foods.

All samples were poured into odor-free, disposable 90 mL covered plastic cups (Sweetheart Cup Co., Inc., Owings Mills, MD, USA) for evaluation. Each panelist received ~60 mL of each product. The samples were served from the refrigerator around 30 min before testing.

After two days of orientation, all samples were evaluated by the panelists in two different days: day one the control juices, and 24 h later the AT-juices. All juices were evaluated in triplicate. The order of product evaluation was randomized, and samples were coded with three-digit random numbers. The descriptive attributes, their definitions, and the list of references used for this study corresponded to the ones reported by Koppel and Chambers [14]. The testing room was at 21 ± 1 °C and 55% ± 5% of relative humidity; the illumination was a combination of natural and non-natural (fluorescent) light.

A modified consensus flavor description method, which uses a panel to determine flavor intensities on a numerical scale from 0 (representing "none") to 15 (representing "extremely strong") was used in this study [16,17,18].

2.4. Statistical Analysis

Physico-chemical data was subjected to statistical analysis using SAS® (version 9.2; SAS Institute, Cary, NC, USA) used for analysis of variance and Fisher's Least Significant Differences test (LSD) for post-hoc mean separation. In addition, Partial least square regression (PLS regression map) was conducted using the Unscrambler version 9.7 (Camo Software, Oslo, Norway) with the objective of relating sensory and instrumental data [19].

3. RESULTS AND DISCUSSION

3.1. Physico-Chemical Analyses

Total soluble solids (TSS), titratable acidity (TA), and maturity indexes (MI) were significantly different ($p \leq 0.05$) among the control juices; however, treatment of pomegranate juices with albedo extract had no significant effects on any of these physico-chemical parameters in any sample ($p > 0.05$ between each control and its corresponding AT-juice) (Table 1).

Vázquez-Araújo *et al.* [15] reported that the MI of some blended juices, which main ingredient was pomegranate juice, was related to consumer overall liking. Calín-Sanchez *et al.* [11] reported a similar relationship between MI and consumer liking, when studying pure fresh pomegranate juice. In addition, MI and has been commonly used as an index of sensory acceptability in different juices [20,21,22]. Consequently and because treatment with albedo extract did not change TSS, it should not influence the acceptability of the juices under study.

Table 1. Physico-chemical characteristics (total soluble solids content: TSS, titratable acidity: TA, maturity index: MI, and total polyphenol content: TPC) of original and AT-juices.

		Physico-chemical characteristics †			
Sample	Cluster ‡	TSS (°Brix)	TA (g citric acid L^{-1})	MI	TPC (mg gallic acid eq. L^{-1})
		Before albedo extract addition			
324	2	16.6 ± 0.3 bcd	13.2 ± 0.1 de	12.6 ± 0.4 c	2577 ± 50 cd
981	5	18.2 ± 0.5 a	10.0 ± 0.2 def	19.6 ± 0.8 b	997 ± 77 f
618	1	13.7 ± 0.4 f	5.13 ± 0.1 f	27.3 ± 2.4 a	2003 ± 70 e
612	4	15.4 ± 0.3 de	16.0 ± 0.3 bc	10.2 ± 1.5 c	846 ± 4 f
707	3	17.6 ± 0.9 abc	20.9 ± 0.2 a	8.5 ± 3.3 c	3784 ± 5 b
		After 24 h of maceration with albedo extract			
324	2	16.2 ± 0.1 cde	14.0 ± 0.1 cd	11.7 ± 0.3 c	3732 ± 116 b
981	5	18.0 ± 0.1 ab	9.5 ± 0.1 ef	19.4 ± 0.1 b	2163 ± 113 de
618	1	13.7 ± 0.1 f	5.13 ± 0.1 f	27.1 ± 2.1 a	2937 ± 113 c
612	4	14.9 ± 0.2 ef	13.5 ± 0.2 cd	11.3 ± 0.2 c	2367 ± 114 de
707	3	16.6 ± 1.1 bcd	19.4 ± 0.1 ab	8.5 ± 0.7 c	5072 ± 233 a

† Mean of 3 replications. Values followed by the different letter, in the same column, were significantly different ($p < 0.05$), Fisher's Least Significant Difference (LSD). ‡ According to Koppel and Chambers [14].

Considering the MIs of the different juices, 618 might be the juice with the highest acceptability of the studied samples, because it had the highest MI ($p \leq 0.05$). However, and although this could be a reasonable assumption, consumer studies should be conducted to prove it. This statement can be affected by other factors, such as the content of citric acid. Hasnaoui *et al.* [23] reported that citric acid content controls pomegranate sourness and a low content of this organic sugars leads to intense sweetness perception.

Dafny-Yalin *et al.* [24] studied the main differences among juices prepared from arils and from pomegranate peel homogenates and found that the later exhibited lower levels of TSS, TA, soluble sugars and organic acids than aril juices. Different manufacturing processes [25] and/or different pomegranate cultivars [10,11], brought different TSS, TAs, and MIs to the original samples in the present study; however, the addition of only a 10% of albedo was not enough to have a significant effect in these parameters.

Total phenolic content (TPC) and antioxidant activity (AA) have been studied by different authors in pomegranate and pomegranate-based products from different countries: e.g., Spain [11], and Iran [10]. As can be seen in Table 1, control samples had different differences in their TPC values. No information about the pomegranate cultivar was provided in the samples labeling, but these differences may be due to geographical origin, cultivar, and/or different manufacturing procedures. Adding albedo homogenate increased TPC in the juices 1.3–2.7 times. The juices which had an initial lower TPC (samples 981 and 612) were the samples which experimented higher increases, rising values over 2100 mg gallic acid equivalents L^{-1}. Consequently, maceration of juices with albedo extract could be an interesting option to develop more competitive and healthy products. Sample 707, which had already the highest TPC (~3700 mg gallic acid equivalents L^{-1}), rise to more than 5000 mg gallic acid equivalents L^{-1} after treatment. Vrhovsek *et al.* [26] stated a recommended daily intake (RDI) of polyphenols of 1 g day^{-1}, so the consumption of only 200 mL of this AT-juice will meet the RDI for polyphenols.

3.2. Volatile Composition and Sensory Analyses

Table 2 shows the main differences in the volatile profile among all studied samples. Up to 69 compounds belonging to 9 chemical families were detected in the juices: alcohols, aldehydes, ketones, acids, furans, esters, benzene derivatives, terpene derivatives, and lactones. All these groups, but lactones, had been reported in commercial pomegranate juices by Vázquez-Araújo *et al.* [27]. Sample 981 was the juice with the highest concentration of volatile compounds (Table 2); esters, benzene derivatives, and terpenes predominated in this sample. Esters are significant aromatic compounds for fruits, synthesized only by intact cells during the β-oxidation of fatty acids or from amino acid metabolism [28], but have been reported previously in pomegranate fresh juices [11,27,29], but in much lower contents, especially in the headspace of the juices. Due to the high concentration of these compounds and benzene derivatives, it could be assumed that were used as flavorings to increase the overall aroma of the juice. Pomegranate juice has low concentration of volatile compounds, which leads to low intensities in odor and aroma [11]. Sample 981 was the only concentrate, so during its production there was a concentration stage in which the volatile compounds would be lost and replaced, or collected and after that added back to the juice.

Table 2. Aromatic volatile compounds found in pomegranate juices.

Code	Compound	RI (Lit.) [*]	RI (Exp.)	Volatile compounds (mg kg^{-1}) [#]									
				324	324 AT	981	981 AT	618	618 AT	612	612 AT	707	707 AT
						Alcohols							
A1	2-Butanol	605	608	n.d.	n.d.	0.004	n.d.	n.d.	n.d.	n.d.	n.d.	n.d.	n.d.
A2	1-Pentanol	746	759	n.d.	n.d.	0.037	0.030	0.002	0.002	n.d.	n.d.	n.d.	n.d.
A3	3-Hexen-1-ol [¥]	860	859	n.d.	n.d.	0.065	n.d.	0.047	n.d.	n.d.	n.d.	n.d.	n.d.
A4	4-Methyl-1-pentanol	872	860	n.d.	0.007	0.050	0.044	0.024	0.020	n.d.	0.005	0.028	0.028
A5	3-Octanol [¥]	997	990	0.005	n.d.	n.d.	n.d.	n.d.	n.d.	n.d.	n.d.	n.d.	n.d.
A6	2-Ethyl-1-hexanol	1030	1032	0.067	0.029	0.027	0.035	0.051	0.040	0.040	0.032	0.044	0.028
A7	Linalool	1101	1098	0.015	n.d.	n.d.	n.d.	0.024	0.034	n.d.	n.d.	n.d.	n.d.
A8	Viridiflorol [‡]	1617	1569	n.d.	n.d.	0.042	0.019	n.d.	n.d.	n.d.	n.d.	n.d.	n.d.
	Total			0.087	0.036	0.226	0.128	0.149	0.097	0.040	0.037	0.071	0.056
						Aldehydes							
A9	2-Pentenal	750	721	n.d.	n.d.	n.d.	n.d.	n.d.	n.d.	0.004	0.003	n.d.	n.d.
A10	2-Methyl-2-butenal	750	741	n.d.	n.d.	n.d.	n.d.	n.d.	n.d.	0.007	0.003	n.d.	n.d.
A11	3-Hexenal [¥]	800	800	n.d.	0.007	n.d.	n.d.	n.d.	0.016	n.d.	n.d.	n.d.	n.d.
A12	Hexanal [¥]	802	802	n.d.	0.060	n.d.	0.047	0.008	0.053	n.d.	0.034	n.d.	0.089
A13	2-Hexenal [¥]	860	865	0.006	0.193	n.d.	0.213	n.d.	0.197	n.d.	0.100	0.116	0.219
A14	Octanal [¥]	1005	1001	0.007	n.d.	n.d.	n.d.	0.006	0.006	0.003	0.004	n.d	n.d.
A15	Nonanal [¥]	1108	1108	0.017	0.014	n.d.	n.d.	0.011	0.018	0.012	0.019	n.d.	0.010

	Total			0.029	0.274	n.d.	0.260	0.025	0.289	0.026	0.164	0.116	0.318
	Ketones												
A16	3-Methyl-2-pentanone	751	750	n.d.	n.d.	0.007	0.006	n.d.	n.d.	n.d.	n.d.	n.d.	n.d.
A17	2-Methyl-2-hepten-6-one ¥	987	986	n.d.	n.d.	n.d.	n.d.	n.d.	0.003	n.d.	n.d.	n.d.	n.d.
A18	β-Damascenone	1400	1384	n.d.	n.d.	n.d.	n.d.	n.d.	n.d.	0.008	n.d.	n.d.	n.d.
A19	β-Ionone	1499	1503	n.d.	n.d.	0.026	0.008	n.d.	n.d.	n.d.	n.d.	n.d.	n.d.
	Total			n.d.	n.d.	0.033	0.014	n.d.	0.003	0.008	n.d.	n.d.	n.d.
	Acids												
A20	Acetic acid †	602	600	0.014	0.003	n.d.	0.010	0.029	0.012	0.134	0.237	0.013	0.006
A21	4-Butoxy butanoic acid ‡	856	856	n.d.	n.d.	n.d.	n.d.	n.d.	n.d.	n.d.	n.d.	0.104	0.038
	Total			0.014	0.003	n.d.	0.010	0.029	0.012	0.134	0.237	0.117	0.044

Values are mean of 3 replications. * [30]. ¥ Aroma compounds found in fresh pomegranate juices [11,27,29]. † Semi-quantification relative to the internal standard concentration. ‡ Tentatively identified: only mass spectral data (retention index, RI, was not found in the literature [30]), the experimental RI differs in more than 20 units from the literature RI, or no standard was available. Standard error was ≤0.01 for all mean values.

Just by looking at the differences in total volatiles between control and AT-juices, no general trend could be deduced, because two of the samples (324 and 612) had higher concentration of total volatile compounds after the albedo homogenate treatment, but the other three samples had the opposite behavior (981, 618, and 612). Only studying chemical families some tendency can be seen: a decrease in the total concentration of alcohols and an increase in the total concentration of aldehydes. Mainly hexanal, 2-hexenal, and 3-hexenal were the compounds which rise seemed to be directly related with the albedo, because they were absent in the original/control juices, but were present in all AT-juices (Table 2). These compounds have sensory descriptors such as fatty, green, apple, floral, or fruity.

Figure 1 shows the main relationships among instrumental and sensory attributes in the different juices. When taking into account the first two dimensions of the PLSR biplot (PLS1 and PLS2), 63% variation in instrumental data explained 58% variation in the sensory data. As can be seeing in the map, most of the differences were related with the samples, and not with treatment. Each one of the AT-sample was close to the original sample and had similar sensory characteristics. Figure 2, Figure 3,Figure 4, Figure 5 and Figure 6 illustrate the different sensory attributes detected in each juice, and how these attributes changed after the maceration with pomegranate albedo. These radar graphs represent the detection of all flavors and mouth-sensations of the juices sequentially: starting from the top of the graph and reading it clockwise. Attributes "sour2", "bitter2", or "astringent2" represented the sourness, bitterness, or astringency of the samples detected on a second time during the testing.

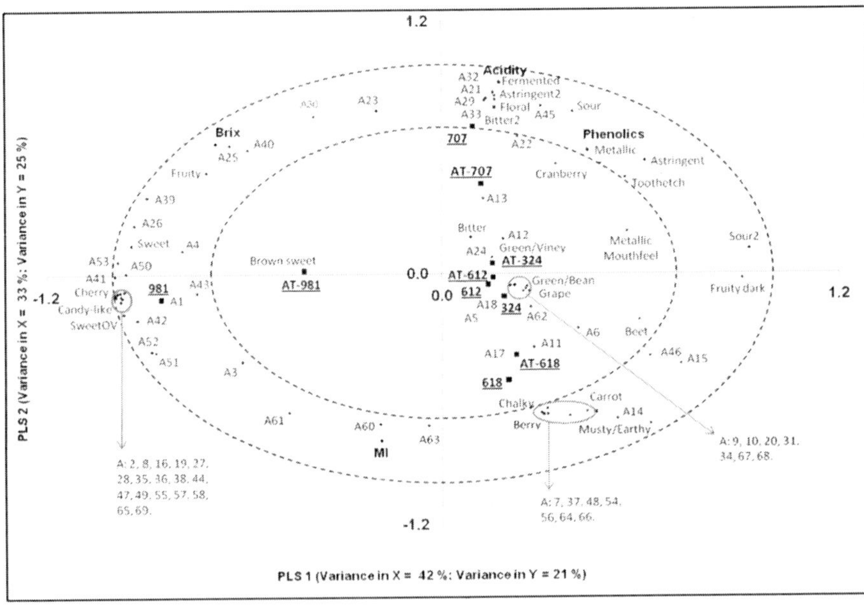

Figure 1. PLS regression map showing the relationship among instrumental and sensory data in all studied juices. *Legend*: Juice samples: indicated in bold and underlined font. Samples with an "AT" before the sample code represent samples "after" albedo treatment. ● Sensory attributes and instrumental volatile compounds. ♦ Physico-chemical data: Brix-TSS, Acidity-TA, MI-MI, Phenolics-TPC.

Samples 324 and 612 had a similar position in Figure 1. These samples had the lower content on total aromatic compounds and similar MI, but different TPC and different sensory profiles. Fifteen different sensory attributes were detected in samples 324 (Figure 2), and only ten in sample 612 (Figure 5). Although the TPC content of the original sample 324 was significantly higher than that of sample 612, the intensities of bitterness and astringency (parameters related with phenolics in literature, e.g., Vardin and Fenercioglu [31], were scored higher in sample 612. These attributes might be slightly masked in sample 324 due to the presence of other flavors that were not present in sample 612, e.g., cranberry or grape. Sourness and astringency have been reported as attributes that dislike consumers in pomegranate fresh juice [11]. In this way, Granato *et al.* [32] reported that pomegranate juices were characterized by high levels of astringency, and concluded that this may hinder their sensory acceptance; however, this problem could be overcome by including a health claim on the label, then the possible beneficial health effects probably drive their consumption. Other options to mask high astringency is to use a mixture of pomegranate cultivars, including sweet pomegranates, such as *Mollar de Elche*; the sweetness of this type of cultivars will help in masking excessive astringency. Sample AT-324 had higher sourness and astringency than original sample 324; these two attributes were not affected in sample AT-612, maybe because they were already high in the original/control juice. These results showed that consumer overall liking of sample 324, and the samples corresponding to cluster 2 [14] may be affected if macerated with albedo, but the effect of the treatment may not affect sample 612 (representative of cluster 4 as reported by Koppel and Chambers [14]). In addition, both AT-samples had new flavor notes that were not present in the original juices: green-viney and green-bean, respectively. These green flavors could to be related with the increase in hexanal (A12) and 2-hexenal (A13) concentrations (Table 2).

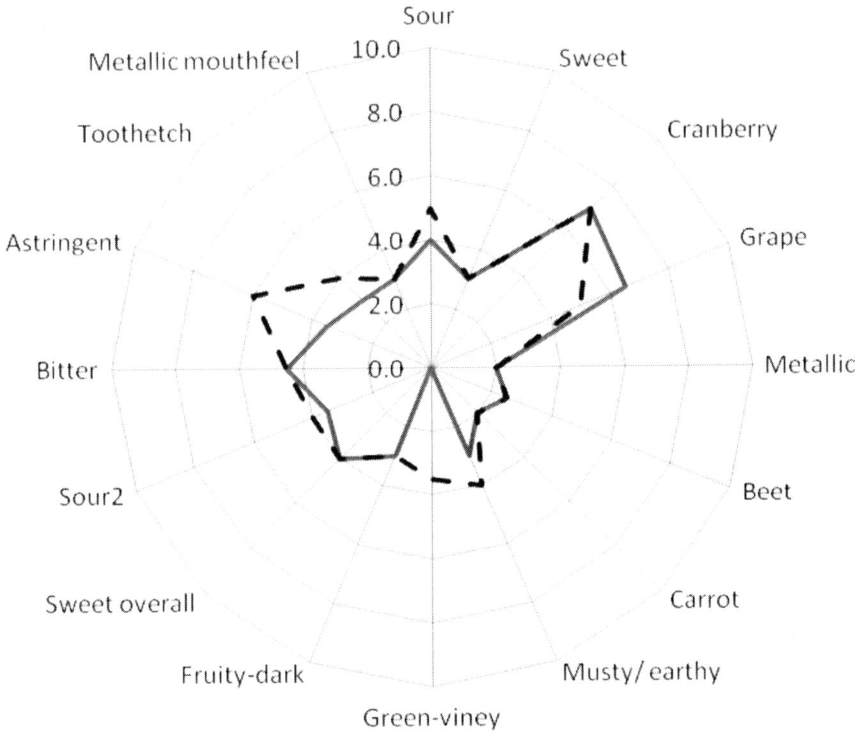

Figure 2. Spider plot showing the differences in the flavor characteristics of sample 324 "before" and "after" albedo treatment. *Legend*: Control juice indicated with a continuous line; AT-juice indicated with a discontinuous line. A numerical scale from 0 (representing "none") to 15 (representing "extremely strong") was used to obtain the data; the upper part of the scale is not shown in the graph. Differences can be considered significant ($p < 0.05$) when the difference between values were higher than 0.5 units.

Samples 981 and AT-981 had the higher concentration in total volatile compounds and also the higher TSS. As discussed previously, most benzene derivatives (A41–A50) were found in these two samples, especially high was the content of benzaldehyde (cherry, bitter almond). In addition, a considerable amount of esters (A22–A38) and terpenes (A51–A68) were found in these samples, making their aroma profile the most complicated (Figure 3). Despite this high amount of volatiles, only ten flavor attributes were detected in these samples (representative of cluster 5 as reported by Koppel and Chambers [14]), characterized by candy-like and cherry attributes, and a high overall sweetness.

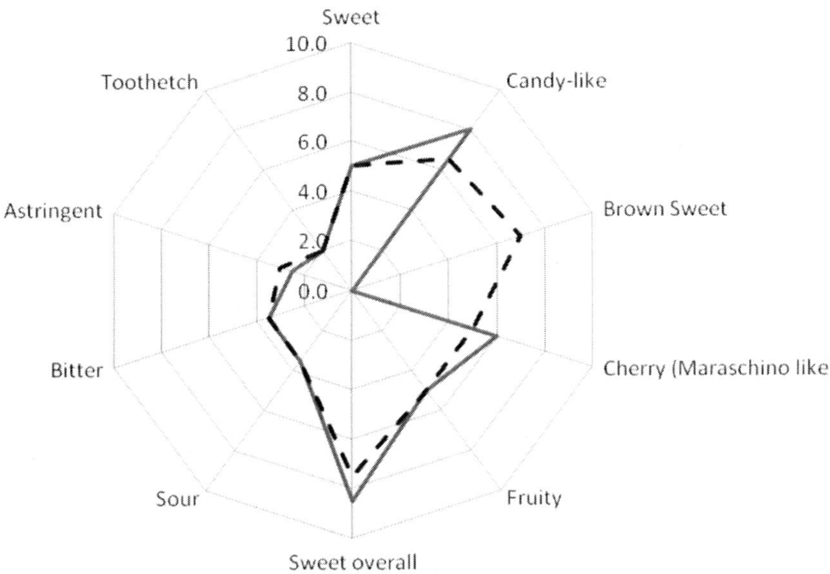

Figure 3. Spider plot showing the differences in the flavor characteristics of sample 981 "before" and "after" albedo treatment.*Legend*: Control juice indicated with a continuous line; AT-juice indicated with a discontinuous line. A numerical scale from 0 (representing "none") to 15 (representing "extremely strong") was used to obtain the data; the upper part of the scale is not shown in the graph. Differences can be considered significant ($p < 0.05$) when the difference between values were higher than 0.5 units.

Samples 618 and AT-618, the ones with the higher MI, were characterized by dark-fruity, musty/earthy and carrot flavors, and also for a chalky mouthfeel (representative of cluster 1 as reported by Koppel and Chambers [14]). As can be seen in Figure 4, bitterness and astringency of these samples were considerably affected by the treatment (maceration with pomegranate albedo). Although some new volatile compounds were detected in the sample AT-618 (e.g., 3-hexenal, 2-hexenal, acetic acid butyl ester, 4-methyl benzaldehyde, α-terpineol), no new flavor notes were found by the panel.

Samples 707 and AT-707 were representative of cluster 3 (fermented flavor and a toothetch mouthfeel; Koppel and Chambers [14]). These samples had also high sourness, bitterness and astringency, persistent attributes which appeared a second time during the testing with a high intensity as well.Figure 6 shows the changes in the sensory profile from sample 707 to sample AT-707. As can be seen, mainly bitterness was affected in this sample, so it is assumable that consumer overall liking may be affected as well. This original sample was characterized for its high TPC, and also for its high bitter and astringent character. These results seemed to confirm the relationship between these sensory attributes and the phenolic compounds (Figure 1), but only when the concentrations are high, and depending on the original flavor profile of the product (as demonstrated when comparing samples 612 and 324).

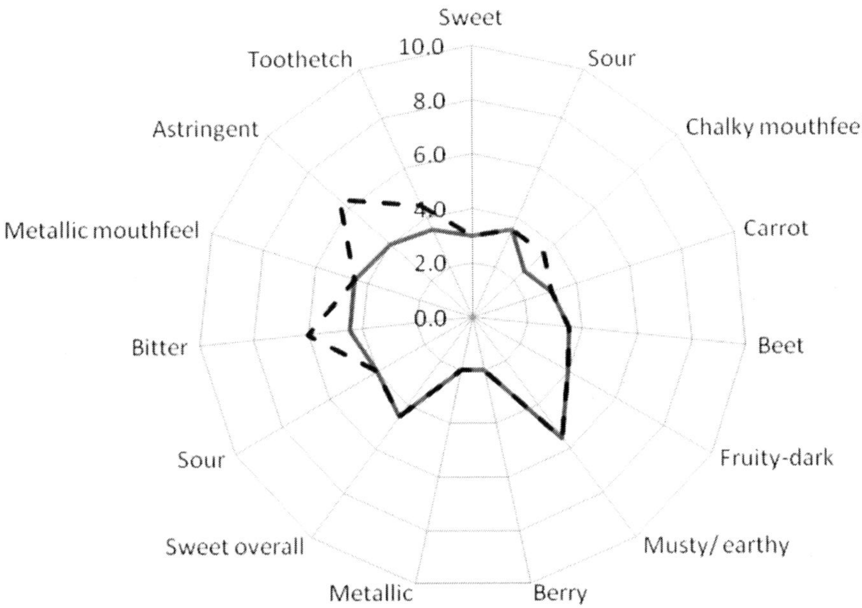

Figure 4. Spider plot showing the differences in the flavor characteristics of sample 618 "before" and "after" albedo treatment.*Legend*: Control juice indicated with a continuous line; AT-juice indicated with a discontinuous line. A numerical scale from 0 (representing "none") to 15 (representing "extremely strong") was used to obtain the data; the upper part of the scale is not shown in the graph. Differences can be considered significant ($p < 0.05$) when the difference between values were higher than 0.5 units.

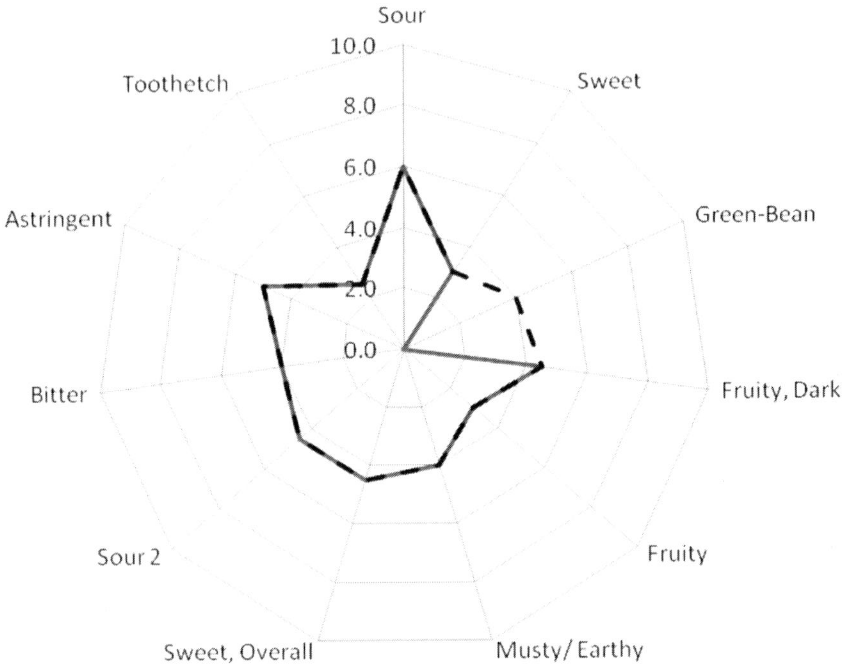

Figure 5. Spider plot showing the differences in the flavor characteristics of sample 612 "before" and "after" albedo treatment.*Legend*: Control juice indicated with a continuous line; AT-juice indicated with a discontinuous line. A numerical scale from 0 (representing "none") to 15 (representing "extremely strong") was used to obtain the data; the upper part of the scale is not shown in the graph. Differences can be considered significant ($p < 0.05$) when the difference between values were higher than 0.5 units.

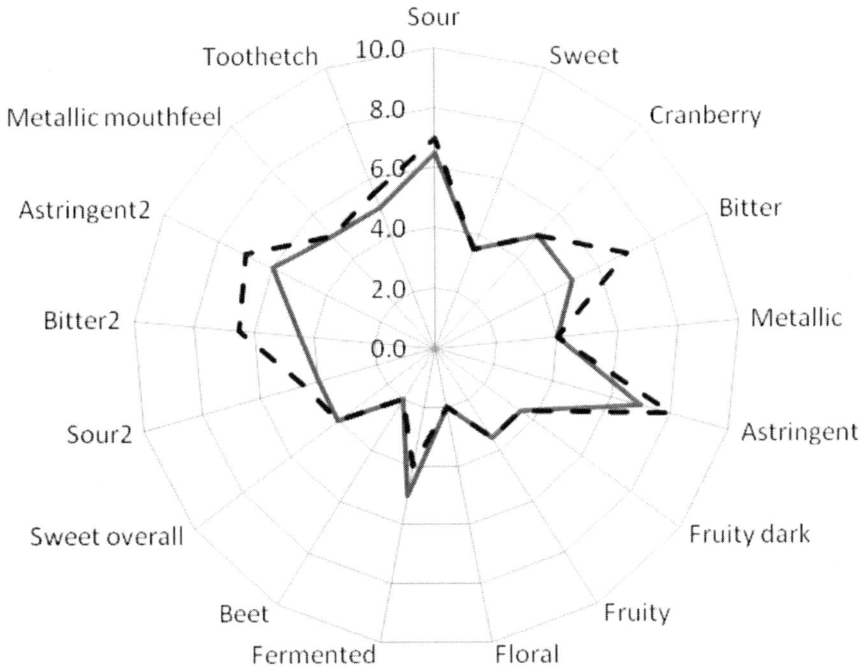

Figure 6. Spider plot showing the differences in the flavor characteristics of sample 707 "before" and "after" albedo treatment.*Legend*: Original juice indicated with a continuous line; AT-juice indicated with a discontinuous line. A numerical scale from 0 (representing "none") to 15 (representing "extremely strong") was used to obtain the data; the upper part of the scale is not shown in the graph. Differences can be considered significant ($p < 0.05$) when the difference between values were higher than 0.5 units.

4. CONCLUSIONS

Maceration of pomegranate juices with 10% pomegranate albedo for 24 h had no significant effects on TSS, TA, and MI in any of the five pomegranate juices. Maceration with albedo significantly increased TPC in all samples, with increases being 1.3-2.7-fold.

Some volatile compounds appeared after the maceration with albedo, for example hexanal, 2-hexenal, and 3-hexenal, which brought green flavor notes to some of the samples (AT-324, and AT-612). Astringency and bitterness of some juices were higher after the maceration, but not in all samples. Samples representing clusters 4 and 5 as described by Koppel and Chambers [14] (musty/earthy and candy-like and sweet overall flavors, respectively) had new flavor notes after albedo addition (green-bean and brown sweet), but sourness, bitterness or astringency were not affected. Maceration with 10% pomegranate albedo seemed to be a good strategy to elaborate healthy and competitive juices, at least to all this companies which products belong to clusters 4 and 5, and have low values of TPC. Consumer studies should be conducted to confirm the impact that the increases in bitterness and/or astringency may have in the acceptance of products from clusters 1, 2 and 3.

AUTHOR CONTRIBUTIONS

L.V-A. and E.C. planned and designed the experiments. L.V-A. performed the experiments. L.V-A. and A.A.C-B. analyzed the data. L.V-A. wrote the manuscript. L.V-A., E.C., and A.A.C-B. edited the manuscript.

CONFLICTS OF INTEREST

The authors declare no conflict of interest.

REFERENCES

1. Basu, A.; Penugonda, K. Pomegranate juice: A heart-healthy fruit juice. *Nutr. Rev.* **2009**, *67*, 49–56.

2. Rettig, M.B.; Heber, D.; An, J.; Seeram, N.P.; Rao, J.Y.; Liu, H.; Klatte, T.; Belldegrun, A.; Moro, A.; Henning, S.M.; *et al.* Pomegranate extract inhibits androgen-independent prostate cancer growth through a nuclear factor-KB-dependent mechanism. *Mol. Cancer Ther.* **2008**, *9*, 2662–2671.

3. Saruwatari, A.; Okamura, S.; Nakajima, Y.; Narukawa, Y.; Takeda, T.; Tamura, H. Pomegranate juice inhibits sulfoconjugation in Caco-2 human colon carcinoma cells. *J. Med. Food* **2008**, *11*, 623–628.

4. Johanningsmeier, S.D.; Harris, G.K. Pomegranate as a functional food and nutraceutical source.*Ann. Rev. Food Sci. Technol.* **2011**, *2*, 181–201.

5. Tezcan, F.; Gültekin-Özgüven, M.; Tuğba, D.; Özçelik, B.; Erim, B.F. Antioxidant activity and total phenolic, organic acid and sugar content in commercial pomegranate juices. *Food Chem.* **2009**, *115*, 873–877.

6. Mousavinejad, G.; Emam-Djomeh, Z.; Rezaei, K.; Khodaparast, M.H.H. Identification and quantification of phenolic compounds and their effects on antioxidant activity in pomegranate juices of eight Iranian cultivars. *Food Chem.* **2009**, *115*, 1274–1278.

7. Borochov-Neori, H.; Judeinstein, S.; Tripler, E.; Harari, M.; Greenberg, A.; Shomer, I.; Holland, D. Seasonal and cultivar variations in antioxidant and sensory quality of pomegranate (*Punica granatum* L.) fruit. *J. Food Compos. Anal.* **2009**, *22*, 189–195.

8. Gil, M.I.; Tomas-Barberán, F.A.; Hess-Pierce, B.; Holcroft, D.M.; Kader, A.A. Antioxidant activity of pomegranate juice and its relationship with phenolic composition and processing. *J. Agric. Food Chem.* **2000**, *48*, 4581–4589.

9. Zang, L.H.; Li, L.L.; Li, Y.X.; Zhang, Y.H. *In vitro* antioxidant activities of fruits and leaves of pomegranate. *Acta Hortic.* **2008**, *765*, 31–34.

10. Tehranifar, A.; Zarei, M.; Nemati, Z.; Esfandiyari, B.; Vazifeshenas, M.R. Investigation of physico-chemical properties and antioxidant activity of twenty Iranian pomegranate (*Punica granatum* L.) cultivars. *Sci. Hort.* **2010**, *126*, 180–185.

11. Calín-Sánchez, A.; Martínez, J.J.; Vázquez-Araújo, L.; Burló, F.; Melgarejo, P.; Carbonell-Barrachina, A.A. Volatile composition and sensory quality of Spanish pomegranates (*Punica granatum* L.). *J. Sci. Food Agric.* **2011**, *91*, 586–592.

12. Ibrahim, M.I. Efficiency of pomegranate peel extract as antimicrobial, antioxidant and protective agents. *World J. Agric. Sci.* **2010**, *6*, 338–344.

13. Panichayupakaranant, P.; Tewtrakul, S.; Yuenyongsawad, S. Antibacterial, anti-inflammatory and anti-allergic activities of standardized pomegranate rind extract. *Food Chem.* **2010**, *123*, 400–403.

14. Koppel, K.; Chambers IV, E. Lexicon to describe appearance and flavor of pomegranate juice. *J. Sens. Stud.* **2010**, *25*, 819–837.

15. Vázquez-Araújo, L.; Chambers, E., IV; Adhikari, K.; Carbonell-Barrachina, A.A. Sensory and physico-chemical characterization of juices made with pomegranate and blueberries, blackberries, or raspberries. *J. Food Sci.* **2010**, *75*, 398–404.

16. Talavera-Bianchi, M.; Chambers, E., IV; Chambers, D. Lexicon to describe flavor fresh leafy vegetables. *J. Sens. Stud.* **2010**, *25*, 163–183.

17. Lee, J.; Chambers, D. Flavors of green tea change little during storage. *J. Sens. Stud.* **2010**, *25*, 512–520.

18. Vázquez-Araújo, L.; Nuncio-Jáuregui, P.N.; Cherdchu, P.; Hernández, F.; Chambers, E., IV; Carbonell-Barrachina, A.A. Physicochemical and descriptive sensory characterization of Spanish pomegranates: Aptitudes for processing and fresh consumption. *Int. J. Food Sci. Technol.* **2014**, *49*, 1663–1672.

19. Granato, D.; Araújo Calado, V.M.; Jarvis, B. Observations on the use of statistical methods in food science and technology. *Food Res. Int.* **2014**, *55*, 137–149.

20. Jordan, R.B.; Seelye, R.J.; McGlone, V.A. A sensory-based alternative to Brix-acid ratio. *Food Technol.* **2001**, *55*, 36–44.

21. Jaya, S.; Das, H. Sensory evaluation of mango drinks using fuzzy logic. *J. Sens. Stud.* **2003**, *18*, 163–176.

22. Oblenland, D.; Collin, S.; Mackey, B.; Sievert, J.; Fjeld, J.; Arpais, M.L. Determinants of flavor acceptability during the maturation of navel oranges. *Postharvest Biol. Technol.* **2009**, *52*, 156–163.

23. Hasnaoui, N.; Mars, M.; Ghaffari, A.; Trifi, M.; Melgarejo, P.; Hernández, F. Seed and juice characterization of pomegranate fruits from Tunisia: Comparison between sour and sweet cultivars revealed interesting properties for prospective industrial application. *Ind. Crop Prod.***2011**, *33*, 374–381.

24. Dafny-Yalin, M.; Glazer, I.; Bar-Ilan, I.; Kerem, Z.; Holland, D.; Amir, R. Color, sugars and organic acids composition in aril juices and peel homogenates prepared from different pomegranate accessions. *J. Agric. Food Chem.* **2010**, *58*, 4342–4352.

25. Nuncio-Jáuregui, N.; Calín-Sánchez, A.; Vázquez-Araújo, L.; Pérez-López, A.J.; Frutos-Fernández, M.J.; Carbonell-Barrachina, A.A. Processing pomegranates for juice and impact on bioactive components. In *Processing and Impact on Active Components in Food*; Academic Press: London, UK, 2014; pp. 629–636.

26. Vrhovsek, U.; Rigo, A.; Tonon, D.; Mattivi, F. Quantitation of polyphenols in different apple varieties. *J. Agric. Food Chem.* **2004**, *52*, 6532–6538.

27. Vázquez-Araújo, L.; Koppel, K.; Chambers, E., IV; Adhikari, K.; Carbonell-Barrachina, A.A. Instrumental and sensory aroma profile of pomegranate juices from the USA: Differences between fresh and commercial juice. *Flavour Frag. J.* **2011**, *26*, 129–138.

28. Belitz, H.D.; Grosch, W.; Schieberle, P. *Food Chemistry*, 4th ed.; Springer: Berlin, Germany, 2009.

29. Melgarejo, P.; Calín-Sánchez, A.; Vázquez-Araújo, L.; Hernández, F.; Martínez, J.J.; Legua, P.; Carbonell-Barrachina, A.A. Volatile composition of pomegranates from 9 Spanish cultivars using headspace solid phase microextraction. *J. Food Sci.* **2011**, *76*, 114–120.

30. NIST (National Institute of Standards and Technology). Available online: http://webbook.nist.gov/chemistry/name-ser.html (accessed on 1 December 2014).

31. Vardin, H.; Fenercioglu, H. Study of the development of pomegranate juice processing technology: Clarification of pomegranate juice. *Nahrung/Food* **2003**, *47*, 300–303.

32. Granato, D.; Karnopp, A.R.; van Ruth, S.M. Characterization and comparison of phenolic composition, antioxidant capacity and instrumental taste profile of juices from different botanical origins. *J. Sci. Food Agric.* 2015.

11

CHAPTER

Consumer Evaluation of Processing Variants of Pomegranate Juice

Erika Anderson, Kadri Koppel * and Edgar Chambers IV

The Sensory Analysis Center, Kansas State University, 1310 Research Park Drive, Manhattan, KS 66502, USA

ABSTRACT

Increasing evidence of their health benefits has boosted the popularity of pomegranates. The effects of processing (e.g., pasteurization, drying) on pomegranate juice characteristics (e.g., color, phenolic content) and sensory attributes have been studied by several authors. The objectives of this study were to (1) understand if processing, such as pasteurization or drying, has an effect on pomegranate juice acceptance, and (2) if acceptance is related to healthy eating habits or variety seeking tendencies. Arils were removed from fresh *Wonderful* pomegranates for juicing or drying. Four treatments were prepared: fresh, fresh frozen, pasteurized, and reconstituted juice from dried arils. Fresh frozen, pasteurized, and reconstituted juices were evaluated by consumers for acceptance. Cluster analysis was conducted and four consumer clusters were found from evaluation of these juice treatments. Each juice was individually disliked by one of three clusters, demonstrating the effect of processing on acceptance. The fourth and largest cluster liked all three treatments. In addition, the consumers were asked to fill in Stage of Change and Variety Seeking scales. Liking scores were not found to be highly associated with healthy eating habits or variety seeking tendencies. This information is beneficial for the fruit processing industry, showing that processing can influence consumer acceptance.

Keywords: pomegranate juice; processing; flavor; consumer acceptance

1. INTRODUCTION

Pomegranates, originally from Iran and the surrounding area, have spread to China, the United States, and Mediterranean countries such as Spain and Morocco. The increase in popularity of pomegranate products within the United States is largely due to health benefits [1]. Cardiovascular, anti-inflammatory, anticancer, and antidiabetic effects encompass the areas of successful *in vitro* and *in vivo* investigation. Although many positive results have been encountered, these studies would benefit from further clinical investigation [2,3,4]. Recent clinical trials of pomegranate juice for hemodialysis patients have shown improvements in systolic blood pressure, pulse pressure, and lipid profiles [5]. Type 2 diabetes patients have also benefited from pomegranate juice consumption by decreasing levels of inflammatory markers that play a role in cardiovascular disease [6]. Polyphenolic compounds including phenolic acids, flavonoids (e.g., anthocyanins), and tannins are largely responsible for these types of health benefits [7].

Many studies have reported processing, such as pasteurization, clarification, concentration, and dehydration effects on health beneficial phenolic compounds. The effect of pasteurization has been shown to increase or decrease anthocyanin content based on pomegranate cultivar and anthocyanin type [8,9]. Clarification, using chemical agents and filtration processes, is used to remove cloud from juices to improve color, flavor, and storage stability. Clarifying methods have displayed differences in levels of phenolic removal, for example gelatin has shown a greater level of total phenolic reduction while polyvinylypolypyrrolidone (PVPP) had a greater effect on anthocyanin reduction [10]. However, filtration processes, such as microfiltration, have shown a greater reduction in anthocyanin content than the use of chemical agents [11].

Reduction of available water through concentration or drying methods not only helps preserve this seasonal fruit, but also increases storage time and provides a more convenient form for shipping. Concentration methods, although resulting in anthocyanin loss, show better retention if microwave heating at high pressures is available [12]. Pomegranate powder production parameters, such as drying agent type and level as well as temperature, can also influence phenolic content. Huruz *et al.* [13] found anthocyanin and phenolic content to be best preserved using higher levels of both maltodextrin and temperatures during drying. However, Robert *et al.* [14] found that soy protein isolates prevent reduction of these compounds better than maltodextrin. Pomegranate arils are commonly dried in India for use in medicines and culinary preparations [15,16]. Singh and Sethi [16] found when drying pomegranate that the anthocyanin level concentrated resulting in a product with a higher level of anthocyanins by weight compared to fresh arils. Jaiswal *et al.* [17] further studied drying of arils and found that drying reduced the level of anthocyanins compared to the levels found in fresh arils. Anthocyanin reductions were greater during sun drying than oven drying.

The understanding of the effects of processing on bioactive compounds is fairly comprehensive; little investigation has taken place to understand the effects these methods have on flavor. Sensory studies have previously investigated descriptive flavor profiles of different pomegranate cultivars [18,19], pomegranate juice blends [20], and commercial products [21,22]. Consumer studies have been used to further understand the liking of differing profiles [19,20,23].

Different scales can be used in addition to acceptance questions to gain further understanding of consumers' behaviors and attitudes [24]. Understanding their interest in health related products could be assessed through methods such as food diaries, food recall, and food frequency questionnaires which can be lengthy and expensive approaches. However, three questions suggested by Soontrunnarudrungsri [25] can be used to segment consumers interested in healthy eating. The variety seeking scale (VARSEEK) may be used to assess a consumer's intrinsic desire to try different food products [26].

Understanding the effects processing imparts on flavor will provide information that is useful in ingredient selection. Additionally, gaining understanding of consumer liking will provide information to processors on the impact of processing methods on acceptability. If dissimilarities are apparent across processing methods, flavor profiles and consumer acceptance should be taken into consideration when selecting products for use. The objectives of this study were to (1) understand the effect of processing including freezing, batch pasteurization, and rehydration of ground, dried arils on consumer acceptability of pomegranate juices, and (2) understand if associations exist between consumer acceptability of these juices and interest in healthy eating and variety seeking tendencies.

2. EXPERIMENTAL SECTION

2.1. Pomegranate Juice Treatments

Fresh pomegranates of the *Wonderful* cultivar grown in the United States were acquired from Youngstown Grape Distributors

(Reedley, CA, USA). The fruits were randomly divided into four groups, washed, and damaged fruits were discarded (Figure 1). Each fruit was cut in half and arils were removed manually by placing each half on top of a Seed-Out (Pahrump, NV, USA), a round raised platform with spokes to support pomegranate halves. Each half was gently tapped and arils falling below Seed-Out were removed for further processing. Any remaining arils connected to the rind were separated manually. Arils were rinsed to further remove any remaining carpellar membranes.

The arils were divided into four parts. Three of the four parts of the resultant pomegranate arils were used for juicing while the fourth was used for drying. Arils used to prepare juice were placed into a Hamilton Beach Health Smart Juice Extractor 67800 (Hamilton Beach/Proctor Silex Inc., South Pines, NC, USA). Approximately 4 L of fresh juice was produced in total over a period of 3 days for immediate analysis. Approximately 17 L was frozen fresh in 946 mL Ziploc freezer bags (SC Johnson, Racine, WI, USA) in a Standex BCF93558-0DX6 blast freezer (Standex, Salem, NH, USA) at (−26 ± 1 °C), for a minimum of 30 min. Once frozen the treatment was transferred to a freezer for storage at −18 ± 1 °C. Of the remaining juice, approximately 15 L was batch pasteurized using a Pastomaster 60 RTX (Carpigiani, Bologna, Italy) at 63 °C for 30 min prior to freezing using the process described for the fresh frozen treatment. The remaining arils in the amount of 29 kg, were dried in batches. Each batch was placed in single layers onto trays of an Excalibur 3926T Food Dehydrator (Excalibur Products, Sacramento, CA, USA). Each batch of arils was dried at 57 °C for 10 h resulting in approximately 6 kg of dried arils. Dried arils were stored frozen in Ziploc freezer bags at −18 ± 1 °C until sample preparation.

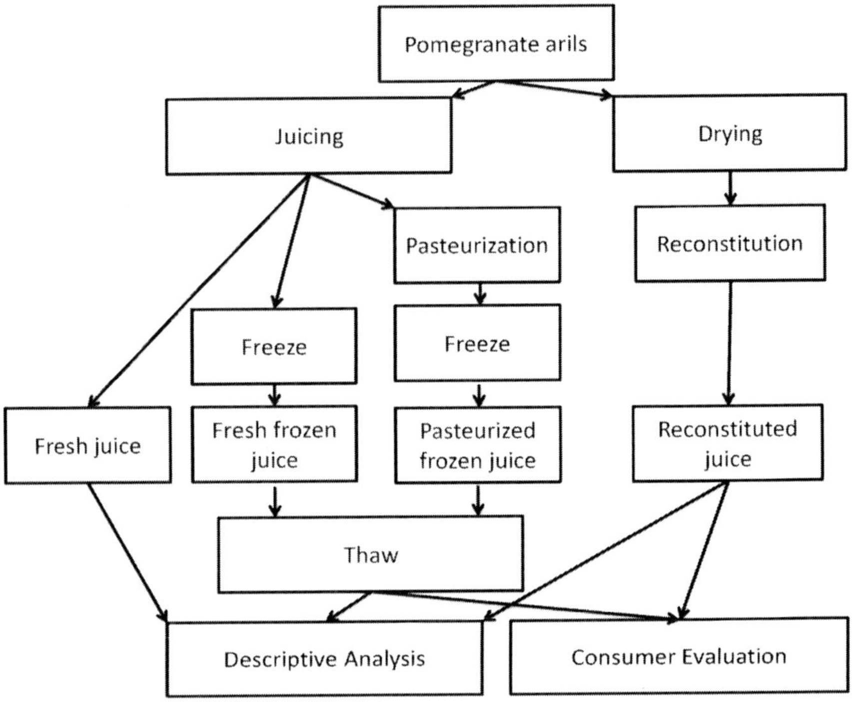

Figure 1. Pomegranate juice processing.

2.2. Descriptive Analysis

The four juice treatments were evaluated using a consensus modified flavor profile previously used by [21,27,28]. Samples were evaluated using descriptive attributes, definitions, and references modified from Koppel and Chambers [21]. A total of 27 attributes were evaluated on the ballot: apple, berry, cranberry, cherry, grape, floral, fruity, carrot, beet, brown spice, fermented, fruity-dark, green-viney, musty/earthy, brown sweet, honey, molasses, overall sweet, sweet, woody, salty, sour, bitter, metallic, astringent, toothetch, chalky mouthfeel.

Each panelist evaluated all attributes for each treatment individually on an intensity scale from 0 (none) to 15 (extremely strong). Following individual evaluation the group discussed each attribute and assigned it an intensity based on reference intensities provided. Treatments were evaluated during 1.5 h sessions, 2 treatment samples per session. Treatment samples were labeled using three-digit codes and served in a random order. Purified water, unsalted crackers, and low-moisture part-skim mozzarella were used as palate cleansers.

Six highly trained panelists with over 1000 h testing experience from the Sensory Analysis Center, Kansas State University (Manhattan, KS, USA) evaluated the treatments. The panelists had prior experience evaluating a variety of products including pomegranates and pomegranate juices. All panelists participated in orientation of the juices they were testing. Thirty milliliters of treatment was served in an odor-free, disposable 96 mL cup at ambient room temperature (21 ± 1 °C). Additional treatment samples were available upon request.

2.3. Consumer Evaluation

2.3.1. Sample Preparation

Frozen juice samples were placed in the refrigerator at 4 °C two days prior to the study to facilitate thawing. Frozen, dried arils were ground using a Black and Decker Smart Grind CBG5 (Applica Consumer Products Inc., Miami Lakes, FL, USA). The ground arils were rehydrated with purified water using a ratio of 22.2 parts arils to 78.2 parts water. The slurry of ground arils was allowed to rest for 10 min to reach a level of 14–15 degrees brix. After resting, the solids were separated from the resulting juice by passing through a sieve two times and the brix measurement was confirmed using a MT-032 Hand Refractometer (QA Supplies Perishable and Environmental Management, Norfolk, VA, USA). A total of 30 mL of juice were poured into odor-free, disposable 96 mL cups and stored in the refrigerator at 4 °C prior to evaluation. Treatment samples were served at 5 ± 1 °C.

2.3.2. Consumers

Consumers were recruited from the Manhattan, Kansas area to participate in pomegranate juice evaluation. Initial screening was performed using Compusense at-hand® (Compusense Inc., Guelph, Ontario, Canada). The 100 consumers that participated met all eligibility requirements including: had no food allergies, drank juice a minimum of 2 to 3 times a week and were willing to try pomegranate juice. All consumers were scheduled by telephone and received a confirmation email of the scheduled participation time. In total, 45% males and 55% females participated in the consumer study (Table 1).

Table 1. Consumer demographics.

Gender	Consumer %	Age Range	Consumer %
Female	55	18–24	14
Male	45	25–34	19
		35–44	15
		45–54	21
		55–64	29

2.3.3. Test Design and Sample Evaluation

Compusense at-hand® (Compusense Inc., Guelph, Ontario, Canada) was used to generate the test design that was balanced for carryover effects. All positions and orders were blocked by groups of six for a possible 120 consumers.

Evaluation of treatments was completed in a total of six sessions over a two-day period; each consumer participated in one of the six sessions. Before beginning the session, consumers were provided with information regarding the study and the procedures that would be used during the testing session. The consumers then electronically signed consent forms stating that they agreed to participate in the study and could withdraw at any time without penalty. Each session lasted for approximately 30 min. Consumers were provided with purified water and unsalted crackers for use in between treatment samples, and optional cups for expectoration.

Compusense at-hand® (Compusense Inc., Guelph, Ontario, Canada) was used on iPads to collect data. Panelists first confirmed their demographic information. Then treatment samples were presented monadically according to each panelists serving order. The consumers rated each sample using five categories of liking including: overall, fruit flavor, sweetness, sourness, and aftertaste. A nine-point hedonic scale ranging from dislike extremely to like extremely was used to indicate level of liking.

After completing sample evaluation, consumers rated their association with the eight statements on the VARSEEK scale using a five point scale; ranging from completely disagree (1) to completely agree (5) [26]. The consumers were grouped based on their scores into three groups: a low (score ≤25), medium (26–34), and high (≥35). The scale consisted of the following statements: (1) When I eat out, I like to try the most unusual items, even if I am not sure I would like them, (2) While preparing foods or snacks, I like to try out new recipes, (3) I think it is fun to try out food items one is not familiar with, (4) I am eager to know what kinds of foods people from other countries eat, (5) I like to eat exotic foods, (6) Items on the menu that I am unfamiliar with make me curious, (7) I prefer to eat food products that I am used to, and 8) I am curious about food products I am not familiar with.

Additionally, each consumer answered three questions described by Soontrunnarudrungsri [25] regarding their habits towards a healthy diet: (1) Do you think you usually eat a healthy diet overall? (options included: yes/no), (2) Do you plan to change your diet to be healthier? (options included: Yes, I am considering changing my overall diet in the next month to be more healthful; Yes, I am considering changing my overall diet in the next 6 months to be more healthful; No, I have no plans to change my overall diet to increase its healthfulness), and (3) How long ago did you change and start eating a healthier diet overall? (options included: I made serious changes in my overall diet to be more healthful in the past 6 months; I made changes to a more healthful diet and have eaten that way for 6 months or more). The first question asks the consumer if they regularly consume a healthy diet. Upon answering "no" the consumer is then asked if they plan to change their diet to be healthier. If not planning to change their diet they were considered to be in the Precontemplation stage. If planning to change in the next month to 6 months they were considered to be in the Preparation and Contemplation stages respectively. If answering "yes" to the first question and making serious changes in the last 6 months they were considered in the Action stage. If indicating a healthy diet for longer than 6 months they were considered in the Maintenance stage [25].

After the session was completed, consumers were monetarily compensated for their time to participate in the study.

2.3.4. Statistical Analysis

Analysis of Variance (ANOVA) was used to test if significant differences existed at a 5% significance level between juice treatment liking scores. Ward's minimum variance procedure was performed to determine if distinct groups existed among consumers. ANOVA with post-hoc means separation using Fisher's protected LSD was used to evaluate significant differences at the 5% significance level across all treatments' liking scores by variety seeking level, across all juice liking scores by healthy eating interest, and across all juice liking scores and variety seeking scores for all clusters. All analyses were performed using SAS® statistical software (version 9.3, SAS Institute Inc., Cary, NC, USA).

3. RESULTS AND DISCUSSION

3.1. Results

3.1.1. Descriptive Evaluation

The descriptive analysis results were described by Koppel *et al.* [29]. Briefly, it was found that pomegranate juice samples did not differ as much in their magnitude of each attribute as they did in the number of attributes (Table 2). The fresh juice did not display many fluctuations in comparison to the fresh frozen juice. The most noticeable difference in fresh juice was the lack of carrot flavor, apparent in the other treatments. Although musty/earthy and molasses flavors did not occur in the fresh frozen juice, these flavors were present in the pasteurized and reconstituted juice. The pasteurized juice, in comparison to the fresh frozen, lacked apple flavor, was lower in cherry flavor and astringency, and higher in grape, dark fruity, and brown sweet flavors intensities. The reconstituted juice was the most differentiated showing slight characteristics of brown spice and fermented with lack of beet, green-viney, and honey flavors.

3.1.2. Consumer Acceptance and Clusters

According to the average liking scores for each category assessed, no significant differences were present at a 5% significance level (Figure 2). Further assessment of liking scores was performed to determine if liking clusters existed. Total of four clusters were discovered (Figure 3). Consumer comments overall reflected their cluster preference and scores in the different liking categories. The consumers in Cluster 3 did not display a distinct preference for any of the samples. Cluster 1 particularly disliked the fresh frozen juice, Cluster 2 disliked the rehydrated juice, and Cluster 4 disliked the pasteurized juice.

Table 2. Modified consensus flavor profile scores for pomegranate juice samples.

Attribute	Samples			
	F	FF	P	R
Apple	1.5 *	2	0	1.5
Berry	3	2.5	3	3
Cranberry	3	3.5	3.5	3
Cherry	2	3	1.5	2
Grape	2	2	3	2
Floral	3	2.5	3	3
Fruity	4	4	4	3.5
Carrot	0	1.5	2	2
Beet	2	2	2.5	0
Brown Spice	0	0	0	1.5
Fermented	0	0	0	1.5
Fruity-Dark	2.5	2	3	3.5
Green-Viney	1.5	2	1.5	0
Musty/Earthy	1.5	0	2	1.5
Brown Sweet	2	2	3	3
Honey	1.5	2	1.5	0
Molasses	1.5	0	2	2
Sweet, Overall	4	4	4	4.5
Sweet	2.5	2.5	3	3
Woody	2	2.5	2	2.5
Salty	1.5	2.5	1.5	2.5
Sour	3	3.5	3.5	3.5
Bitter	3.5	3.5	3.5	3.5
Metallic	2	2	2.5	1.5
Astringent	4	5	4	4
Toothetch	2.5	2.5	2	2
Chalky Mouthfeel	2	2	2	2.5

F = Fresh, FF = Fresh Frozen, P = Pasteurized, R = Reconstituted. * Scores given on a scale from 0 to 15 with 0.5 increments.

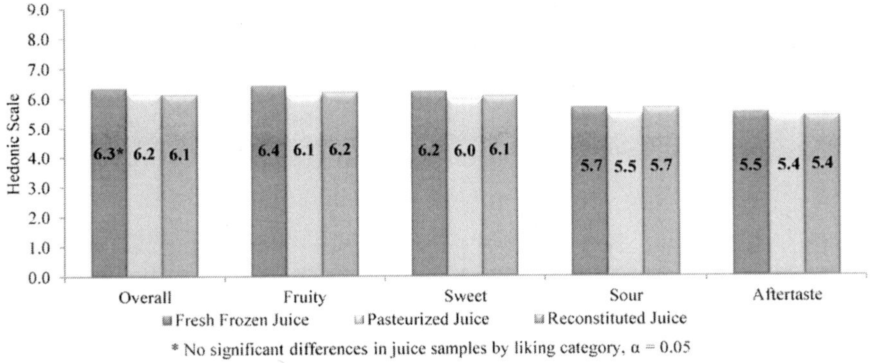

Figure 2. Average consumer liking scores of pomegranate juices. Evaluation on a 9-point hedonic scale, 1—dislike extremely, 9—like extremely.

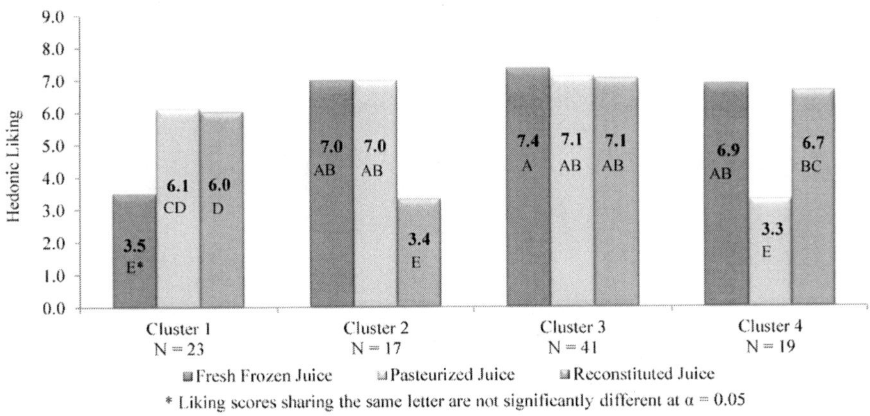

Figure 3. Average overall liking of pomegranate juices by consumer cluster. Evaluation on a 9-point hedonic scale, 1—dislike extremely, 9—like extremely.

The majority of the consumers (80%) felt that they usually eat a healthy diet. The consumers' interest in healthy eating results can be explained by the Stages of Change model [25]. The distribution of consumers throughout the Stages of Change model reflected the results previously found by Soontrunnarudrungsri [25] showing that most of the consumers fall in the Maintenance stage, decreasing in percentage to the Contemplation stage (Table 3). However, no participants in this study were found to be in the Precontemplation stage where they felt their diet was not usually healthy and were not considering changing to a more healthful diet. Overall liking scores showed that the consumers in the Action stage disliked the pasteurized and reconstituted juices significantly less than the consumers in the Contemplation stage liked the pasteurized juice (Figure 4).

Van Trijp and Steenkamp [26] have divided the possible variety seeking scores into the levels of low, medium, and high (Table 4). The average consumer in this study was considered to have a medium intrinsic variety seeking level with a score of 30.33. Based on these category divisions, liking scores were compared across low, medium, and high variety seeking groups. None of the variety seeking groups showed a strong level of liking for any of the juice types (Figure 5). However, compared to medium variety seekers, low variety seekers liked the fresh frozen juice significantly less.

Table 3. Consumer interest in healthy eating habits according to the stages of change questionnaire [25].

Stage of Change	Description	% Consumers
Maintenance	Healthy diet for 6 months or more	61
Action	Made changes in last 6 months to be more healthful	19
Preparation	Considering to a more healthful diet in next month	13
Contemplation	Considering changing to a more healthful diet in the next 6 months	7

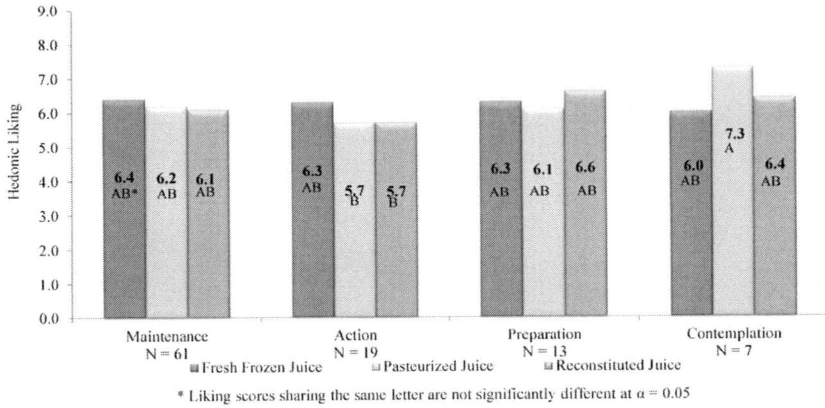

Figure 4. Average overall liking scores by stages of change scale [25]. Evaluation on a 9-point hedonic scale, 1—dislike extremely, 9—like extremely.

Table 4. Consumer variety seeking scores. scores calculated according to VARSEEK scale [26]. Scale 1—disagree completely to 5—agree completely, maximum value of 40.

Variety Seeking Level	Score Range	% Consumers
Low	25 and below	22
Medium	26–34	47
High	35 and above	31

3.2. Discussion

The profiles of the fresh and fresh frozen juices were found to be quite similar which allows for comparison of pasteurized and reconstituted juice to the fresh frozen juice as similar to fresh during the consumer study. Although differences were not large in magnitude, differences did exist among samples for each processing treatment. These differences probably had an influence on consumer acceptability.

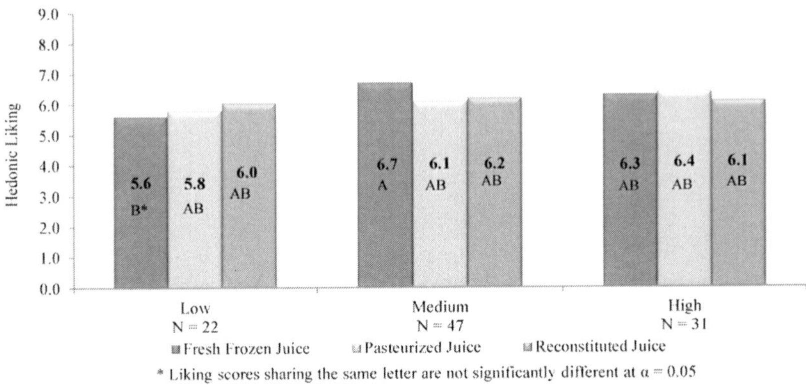

Figure 5. Average overall liking of pomegranate juices by variety seeking category. Evaluation on a 9-point hedonic scale, 1—dislike extremely, 9—like extremely.

Consumer clusters were expected as Koppel *et al.* [23] previously found different liking clusters across countries. The juices in this study were surprisingly well liked, particularly by Cluster 3, as previous studies had not seen as high of liking scores in pomegranate juices or juice blends [23,30,31], and a consumer cluster was expected that did not like any samples as found by Koppel *et al.* [23]. Our finding that groups individually each disliked a treatment indicated that processing methods may play a role in consumer choice when pursuing pomegranate products. From the treatments evaluated, the appearance of specific flavors such as brown spice and fermented and lack of beet, green-viney, and honey may have contributed to the significant dislike of the reconstituted product by Cluster 2. A summation of small changes may have influenced consumer significant dislike of the pasteurized treatment by Cluster 4. These small changes included lack of apple flavor, lower cherry flavor and salty taste, increased grape and musty/earthy flavor.

Processing exhibited changes that may be associated with heating such as increase in level of grape, dark fruity, or brown sweet flavors. Flavors such as these were previously noted in commercial juices that had undergone processing [22]. The elevation of these flavors may have influenced the liking of Cluster 1, which significantly disliked the fresh frozen juice in comparison to those that had undergone processing. Lozano [32] suggested that pasteurization at higher temperatures (such as that greater than 121 °C for 0.1 s) may impart cooked flavor to juice. Further processing may have the capability of enhancing specific notes while decreasing others potentially impacting consumer acceptability. Future studies may address this potential by examining variables such as pasteurization temperature as it relates to product liking.

Stages of change related to interest in healthy eating showed significant differences in acceptance of pomegranate juices in the action and contemplation stages. Examining scores in the other liking categories these differences appear to relate to aftertaste. The pasteurized juice received the highest score in aftertaste liking from the Contemplation group at 6.7. While the pasteurized and reconstituted juices received the lowest aftertaste score from the Action group at 4.5 and 4.8 respectively. These groups showed a trend in aftertaste liking; the Action group was the least accepting of aftertaste, the Maintenance group was the second least accepting, and preparation and Contemplation groups were the most accepting. From this study, we observed that the majority of juice drinkers tend to be more interested in healthy eating and did not discriminate much among juices. Future studies may consider investigating if consumers who are more interested in healthful eating are more discriminatory among other foods and beverages.

VARSEEK scores were expected to be higher with higher liking of samples, as pomegranate juice popularity is relatively recent. We found that acceptability was not well distinguished by VARSEEK scores as there were no significant differences among clusters in this respect and few differences among liking of the treatments at different variety seeking levels. Previously, Lähteenmäki and van

Trijp [33] also were unable to relate variety seeking tendencies to liking in the evaluation of sandwiches, but did find that those with higher variety seeking tendencies tried more "less appropriate" sandwich fillings than those with lower scores. Based on the previous results with sandwiches, it may be useful to consider evaluation of liking scores in combination with juice purchasing habits for a variety of juices to better understand the variety seeking tendencies of pomegranate juice consumers. Moreover, consumer demographics did not demonstrate large differences across clusters, including interest in healthy eating. All clusters had consumers that typically ate a healthy diet overall and if they did not believe they had a healthy diet overall were planning to change their diet to be more healthful.

These findings present additional opportunities for further understanding of sensory effects of pomegranate processing on consumer acceptance. The treatments used in this study were prepared on a small scale. As differences were present in both pasteurization and drying effects, comparison of products produced on a larger scale would be beneficial to understand processing influences in industry. Additionally, processes such as clarification and concentration should be considered for testing as noted changes are produced by polyphenol reduction and heating processes respectively. Both processes are used in industry to improve storage stability. Clarification reduces phenolic compounds decreasing the level of astringency [11]. Methods that involve heating increase the browning index, an indicator of browning reactions such as in Maillard browning or caramelization, forming new flavor compounds [12]. Astringency level fluctuation and flavor changes due to browning may impact consumer acceptance. Lawless *et al.* [20] found that presentation of antioxidant information influenced acceptance of black cherry, Concord grape, and pomegranate juice blends. Several studies have proven that processing has an influence on polyphenols; these influences could be presented to consumers as antioxidant levels to determine their impact on acceptance.

4. CONCLUSIONS

Flavor variations were found as a result of processing methods although small in magnitude. The amount of attributes presenting small changes is influential, presenting distinct differences in consumer liking. Most juice drinkers tend to be interested in healthy eating; however, they did not discriminate much among juices overall. Future investigations may consider examining if this is the case with other product types. Variety seeking tendencies did not show a defined relationship with product liking scores. However, the variation in processing methods exposed three consumer clusters that each individually disliked a particular treatment within the study. The largest cluster surprisingly liked all three juices at a level not previously seen in other studies. Further exploration in the areas of pomegranate processing on consumer acceptance such as full scale processing changes, additional processing methods, and antioxidant information influence would be beneficial to explore.

AUTHOR CONTRIBUTIONS

E.A., K.K., and E.C. planned and designed the experiments; E.A. performed the experiments; E.A., K.K., and E.C. analyzed the data; E.A. wrote the manuscript, K.K. and E.C. edited the manuscript.

CONFLICTS OF INTEREST

The authors declare no conflict of interest.

REFERENCES

1. Hummer, K.E.; Pomper, K.W.; Postman, J.; Graham, C.J.; Stover, E.; Mercure, E.W.; Aradhya, M.; Crisosto, C.H.; Ferguson, L.; Thompson, M.M.; *et al.* Chapter 4: Emerging fruit crops. In *Fruit Breeding*; Badenes, M.L., Byrne, D.H., Eds.; Springer: New York, USA, 2012; Volume 8, pp. 97–147.

2. Faria, A.; Calhau, C. The bioactivity of pomegranate: Impact on health and disease. *Crit. Rev. Food Sci. Nutr.* **2011**, *51*, 626–634.

3. Johanningsmeier, S.D.; Harris, G.K. Pomegranate as a functional food and nutraceutical source.*Annu. Rev. Food Sci. Technol.* **2011**, *2*, 181–201.

4. Viuda-Martos, M.; Fernández-López, J.; Pérez-Álvarez, J.A. Pomegranate and its many functional components as related to human health: A review. *Compr. Rev. Food Sci. Food Saf.* **2010**, *9*, 635–654.

5. Shema-Didi, L.; Kristal, B.; Sela, S.; Geron, R.; Ore, L. Does Pomegranate intake attenuate cardiovascular risk factors in hemodialysis patients? *Nutr. J.* **2014**, *13*, 1–8.

6. Sohrab, G.; Nasrollahzadeh, J.; Zand, H.; Amiri, Z.; Tohidi, M.; Kimiagar, M. Effects of pomegranate juice consumption on inflammatory markers in patients with type 2 diabetes: A randomized, placebo-controlled trial. *J. Res. Med. Sci.* **2014**, *19*, 215–220.

7. Tzulker, R.; Glazer, I.; Bar-Ilan, I.; Holland, D.; Aviram, M.; Amir, R. Antioxidant activity, polyphenol content, and related compounds in different fruit juices and homogenates prepared from 29 different pomegranate accessions. *J. Agric. Food Chem.* **2007**, *55*, 9559–9570.

8. Turfan, Ö.; Türkyılmaz, M.; Yemiş, O.; Özkan, M. Anthocyanin and color changes during processing of pomegranate (*Punica granatum* L., cv. Hicaznar) juice from sacs and whole fruit.*Food Chem.* **2011**, *129*, 1644–1651.

9. Alighourchi, H.; Barzegar, M.; Abbasi, S. Anthocyanins characterization of 15 Iranian pomegranate (*Punica granatum* L.) varieties and their variation after cold storage and pasteurization. *Eur. Food Res. Technol.* **2008**, *227*, 881–887.

10. Vardin, H.; Fenercioğlu, H. Study on the development of pomegranate juice processing technology: Clarification of pomegranate juice. *Nahrung* **2003**, *47*, 300–303.

11. Fischer, U.A.; Dettmann, J.S.; Carle, R.; Kammerer, D.R. Impact of processing and storage on the phenolic profiles and contents of pomegranate (*Punica granatum* L.) juices. *Eur. Food Res. Technol.***2011**, *233*, 797–816.

12. Yousefi, S.; Emam-Djomeh, Z.; Mousavi, S.M.A.; Askari, G.R. Comparing the effects of microwave and conventional heating methods on the evaporation rate and quality attributes of pomegranate (*Punica granatum* L.) juice concentrate. *Food Bioprocess Technol.* **2012**, *5*, 1328–1339.

13. Horuz, E.; Altan, A.; Maskan, M. Spray drying and process optimization of unclarified pomegranate (*Punica granatum*) juice. *Drying Technol.* **2012**, *30*, 787–798.

14. Robert, P.; Gorena, T.; Romero, N.; Sepulveda, E.; Chavez, J.; Saenz, C. Encapsulation of polyphenols and anthocyanins from pomegranate (*Punica granatum*) by spray drying. *Int. J. Food Sci. Technol.* **2010**, *45*, 1386–1394.

15. Kingsly, A.R.P.; Singh, D.B.; Manikantan, M.R.; Jain, R.K. Moisture dependent physical properties of dried

pomegranate seeds (Anardana). *J. Food Eng.* **2006**, *75*, 492–496.

16. Singh, D.; Sethi, V. Screening of pomegranate genotypes for the preparation of quality grade anardana. *J. Food Sci. Technol.* **2003**, *40*, 236–238.

17. Jaiswal, V.; DerMarderosian, A.; Porter, J.R. Anthocyanins and polyphenol oxidase from dried arils of pomegranate (*Punica granatum* L.). *Food Chem.* **2010**, *118*, 11–16.

18. Andreu-Sevilla, A.J.; Mena, P.; Martí, N.; García-Viguera, C.; Carbonell-Barrachina, Á.A. Volatile composition and descriptive sensory analysis of pomegranate juice and wine. *Food Res. Int.* **2013**,*54*, 246–254.

19. Carbonell-Barrachina, A.A.; Calín-Sánchez, A.; Bagatar, B.; Hernández, F.; Legua, P.; Martínez-Font, R.; Melgarejo, P. Potential of Spanish sour-sweet pomegranates (cultivar C25) for the juice industry. *Food Sci. Technol. Int.* **2012**, *18*, 129–138.

20. Lawless, L.J.R.; Threlfall, R.T.; Meullenet, J.; Howard, L.R. Applying a mixture design for consumer optimization of black cherry, concord grape and pomegranate juice blends. *J. Sens. Stud.* **2013**, *28*, 102–112.

21. Koppel, K.; Chambers IV, E. Development and application of a lexicon to describe the flavor of pomegranate juice. *J. Sens. Stud.* **2010**, *25*, 819–837.

22. Vázquez-Araújo, L.; Koppel, K.; Chambers IV, E.; Adhikari, K.; Carbonell-Barrachina, A.A. Instrumental and sensory aroma profile of pomegranate juices from the USA: Differences between fresh and commercial juice. *Flavour Fragrance J.* **2011**, *26*, 129–138.

23. Koppel, K.; Chambers IV, E.; Vázquez-Araújo, L.; Timberg, L.; Carbonell-Barrachina, Á.; Suwonsichon, S. Cross-country comparison of pomegranate juice acceptance in Estonia, Spain, Thailand, and United States. *Food. Qual. Prefer.* **2014**, *31*, 116–123.

24. Lawless, H.T.; Heymann, H. Chapter 14: Acceptance Testing: 14.6 Behavioral and Context-Related Approaches. In *Sensory Evaluation of Food Principles and Practices*; Springer: New York, NY, USA, 2010; pp. 340–343.

25. Soontrunnarudrungsri, A. Development and Validation of Screening Tools for Classification Consumers of Food Products Based on Eating Healthy Criteria. Ph.D. Thesis, Kansas State University, Manhattan, KS, USA, December 2011.

26. Van Trijp, H.C.; Steenkamp, J.E. Consumers' variety seeking tendency with respect to foods: Measurement and managerial implications. *Eur. Rev. Agric. Econ.* **1992**, *19*, 181–195.

27. Cherdchu, P.; Chambers IV, E.; Suwonsichon, T. Sensory lexicon development using trained panelists in Thailand and the USA: Soy sauce. *J. Sens. Stud.* **2013**, *28*, 248–255.

28. Di Donfrancesco, B.; Koppel, K.; Chambers IV, E. An initial lexicon for sensory properties of dry dog food. *J. Sens. Stud.* **2012**, *27*, 498–510.

29. Koppel, K.; Anderson, E.L.; Chambers IV, E. Influence of processing on pomegranate (*Punica granatum* L.) juice flavor and aroma. *J. Sci. Food Agric.* **2014**. in press.

30. Vázquez-Araújo, L.; Chambers IV, E.; Adhikari, K.; Carbonell-Barrachina, Á.A. Sensory and physicochemical characterization of juices made with pomegranate and

blueberries, blackberries, or raspberries. *J. Food Sci.* **2010**, *75*, S398–S404.

31. Endrizzi, I.; Pirretti, G.; Calò, D.G.; Gasperi, F. A consumer study of fresh juices containing berry fruits. *J. Sci. Food Agric.* **2009**, *89*, 1227–1235.

32. Lozano, J.E. Processing of fruits: Elevated temperature, nonthermal, and miscellaneous processing. In *Fruit Manufacturing: Scientific Basis, Engineering Properties, and Deteriorative Reactions of Technological Importance*; Springer: New York, USA, 2006; pp. 55–71.

33. Lähteenmäki, L.; van Trijp, H. Hedonic responses, variety-seeking tendency and expressed variety in sandwich choices. *Appetite* **1995**, *24*, 139–152.

Index